POLYMER SCIENCE AND TECHNOLOGY

AN INTRODUCTION TO GLASS TRANSITION

POLYMER SCIENCE AND TECHNOLOGY

Additional books and e-books in this series can be found
on Nova's website under the Series tab.

POLYMER SCIENCE AND TECHNOLOGY

AN INTRODUCTION TO GLASS TRANSITION

ROBERTA RAMIREZ
EDITOR

nova
science publishers
New York

NOTICE TO THE READER

Library of Congress Cataloging-in-Publication Data

ISBN: 978-1-53615-706-2

Published by Nova Science Publishers, Inc. † New York

CONTENTS

PREFACE

An Introduction to Glass Transition opens with a comparison of entropy function of temperature dependence with configurational entropy, which was published by various authors and found almost the same temperature dependence with overlap. From the dependence of the logarithm of configurational entropy vs. the logarithm of temperature, the authors suggest that it is possible to successfully predict the relations between the values of m for different glass formers.

Following this, microscopic local dynamics were analyzed by way of atomistic molecular dynamics simulations through the conformational transition behavior across a wide range of temperatures. The glass transition temperature may be predicted through the intersection of separate temperature dependences. Such local dynamics were found to become gradually heterogeneous when the temperature went down close to the glassy state.

The closing chapter provides a brief summary of the studies relevant to glass transitions in well-defined lipids systems such as anhydrous and/or water mixed systems. Then, some current problems and future problems are described.

Chapter 1 - The revision of the meaning of the famous Hrubý glass-forming criterion (H) in comparison with other analogous coefficients by e.g., Weinberg (W) or Lu–Liu (LL) is attempting to uncover some generalized correlations between glass-forming ability (GFA) and glass stability (GS). The relative change of the Hrubý parameter is supreme in almost all cases. Relations between the magnitude of the change of the ratios of crystallization (T_c) and melting temperature (T_m) with glass transition temperature (T_g), that are the ratios T_c/T_g and T_m/T_g, determine the order of the values of relative change GS parameters (K) as derivatives dK_H/K_H, dK_W/K_W and dK_{LL}/K_{LL}. The Hrubý parameter is more sensitive in relation to the change of both the super-cooled region and the reduced glass transition temperature. The only exception is the restricted sensitivity respecting some cases of the bulk metallic glasses but is well agreeable for oxide glasses and thus can be commonly employed as a reliable and precise glass forming criterion. The linear correlation of new GS parameters which include fragility and reduced glass transition temperature upon logarithm of cooling rate (R_c) is better than correlation given by K. The stretching exponent increases as a linear function of T/T_g in the interval $1 \le T/T_g < 1.1$ for a given fragility index m. As a result, it follows that the kinetic term in fragility can be neglected. The thermodynamic term, which encompasses a dominant role in fragility, can be determined by the expressions for configurational entropy and configurational heat capacity. The authors compared entropy function of temperature dependence of configurational entropy, which was published by various authors and found almost the same temperature dependence with overlap. From the dependence of the logarithm of configurational entropy vs. the logarithm of temperature, it is possible to successfully predict the relations between the values of m for different glass formers.

Chapter 2 - The old problem of glass transition is regarded as the most important behavior of polymer science, while the fundamental microscopic picture of the polymeric glass-formers is scarcely seen. By way of atomistic molecular dynamics simulations, the microscopic local dynamics were analyzed through the conformational transition behavior across a wide range of temperatures. The glass transition temperature could well be predicted through the intersection of separate temperature dependences. Such local dynamics were found, becoming gradually heterogeneous when the temperature went down close to the glassy state. The frozen torsions were determined through a time dependent fashion, and the statistics related to the frozen fraction and frozen chain length showed that the local dynamics of conformational transition corresponded well with the classical thermodynamic or kinetic theories. The frozen chain segments grew as the temperature decreased in a similar way with that of linear condensation polymerization, and the average frozen chain lengths were related to the configuration entropy during glass transition. The ideal glass transition temperatures, defined in classical thermodynamics theories, were predicted by extrapolating the configuration entropy to zero. The relation between the torsional relaxation time and the configuration entropy showed perfect agreement with the Adam-Gibbs theory around the glass transition temperature. The consecutive frozen torsions had very similar behavior with the cooperative rearranging regions (CRR) defined in the theory. The frozen torsions expanded with a further decrease of temperature until the formation of a volume spanning cluster, which might serve as a premature prototype for the formation of the 'ideal glassy state' with limited accessible configurations.

Chapter 3 - A glass transition is the phenomenon in which an amorphous phase exhibits abrupt changes in derivative thermodynamic properties. Many non-crystalline materials have the potential to exhibit the glass transition, while polymers, inorganic species, and sugars are excellent candidates to maintain a stable glass state under ambient conditions. Recently, investigations involving glass transitions in amphiphilic lipids (or surfactants) consisting of hydrophilic and lipophilic parts such as hydrocarbons have increased. This chapter provides a brief summary of the studies relevant to glass transitions in well-defined lipids systems such as anhydrous and/or water mixed systems. Then, some of the current problems and future tasks to be addressed in the 2020s will be described.

Chapter 1

CHARACTERISTIC TEMPERATURES AND THEIR EXPLOITATION: GLASS TRANSITION, GLASS-FORMING COEFFICIENTS ANALYSIS AND INNOVATIVE CONCEPT OF FRAGILITY

Jaroslav Šesták[1,] and Ana Kozmidis Petrović[2]*

[1]New Technology - Research Center in the Westbohemian Region,
West Bohemian University, Pilsen, Czech Republic
[2]Faculty of Technical Sciences, University of Novi Sad,
Novi Sad, Serbia

ABSTRACT

The revision of the meaning of the famous Hrubý glass-forming criterion (H) in comparison with other analogous coefficients by e.g., Weinberg (W) or Lu–Liu (LL) is attempting to uncover some generalized correlations between glass-forming ability (GFA) and glass stability (GS). The relative change of the Hrubý parameter is supreme in almost all cases. Relations between the magnitude of the change of the ratios of crystallization (T_c) and melting temperature (T_m) with glass transition temperature (T_g), that are the ratios T_c/T_g and T_m/T_g, determine the order of

---
[*] Corresponding Author's E-mail: sestak@fzu.cz.

the values of relative change GS parameters (K) as derivatives dK_H/K_H, dK_W/K_W and dK_{LL}/K_{LL}. The Hrubý parameter is more sensitive in relation to the change of both the super-cooled region and the reduced glass transition temperature. The only exception is the restricted sensitivity respecting some cases of the bulk metallic glasses but is well agreeable for oxide glasses and thus can be commonly employed as a reliable and precise glass forming criterion. The linear correlation of new GS parameters which include fragility and reduced glass transition temperature upon logarithm of cooling rate (R_c) is better than correlation given by K. The stretching exponent increases as a linear function of T/T_g in the interval $1 \leq T/T_g < 1.1$ for a given fragility index m. As a result, it follows that the kinetic term in fragility can be neglected. The thermodynamic term, which encompasses a dominant role in fragility, can be determined by the expressions for configurational entropy and configurational heat capacity. We compared entropy function of temperature dependence of configurational entropy, which was published by various authors and found almost the same temperature dependence with overlap. From the dependence of the logarithm of configurational entropy vs. the logarithm of temperature, it is possible to successfully predict the relations between the values of m for different glass formers.

Keywords: glass transition temperature, glass forming ability, glass stability, glass forming coefficients, kinetic fragility, configurational entropy

MOTO: "The deepest and most interesting unsolved problem in the solid state comprehension is the theory of the nature of glass and of the glass transition."
By Philip W. Anderson (*1923), Nobel Prize winner for Physics 1977

1. INTRODUCTION - A SHORT LOOK INTO HISTORY

Gustav H. J. Tammann (1861-1938), Frederik F. H. Zachariasen (1906-1979), Walter Kauzman (1916-2009), David Turnbull (1915-2007), and Arnošt Hrubý (1919-2014) (see above photo) are most quoted scientists paving basis for a better understanding of glass nature. It is worth noting that the exploration of the associated glass transformation, which is neither of the first nor the second order category, is younger than the theory of radioactivity. Let us rejoin the repeat some historical data [1, 2] given in Turnbull's manuscripts, who in the mid-nineteenth century summed up the essentials saying that glass is often considered to be a solid formed by the continuous hardening of a cooled liquid. By hardening, it is meant an increasing resistance to forces causing the body to flow and permanently change its shape. The quantitative measure of this resistance is the shear viscosity. This property changes continuously upon the transition of a liquid to a glass. It might appear that glasses are simply crystalline solids with extremely small grain sizes; that is, a duplex mixtures of highly ordered and highly disordered material. However, Zachariasen showed long ago that when the atomic co-ordination number is sufficiently small, atoms could be assembled together in a continuous random network, which can properly model a glass. In such a type of structure, local molecular configurations in which the idealized symmetry is non-crystallographic, e.g., pentagonal, are prominent. The density of an assembly of randomly oriented crystallites should decrease with crystallite size, owing to the density deficit at crystallite boundaries, and fall below the density of the continuous random structure at some critical value of the average crystallite size. These structural considerations have led to the concept of an 'ideal' glass state developed for polymeric systems to be situated a solid in internal equilibrium in which there is, just as in a crystalline solid, a definite set of equilibrium positions about which the atoms oscillate. However, in contrast with crystalline solids, the set of equilibrium positions in an ideal glass does not exhibit translational symmetry; that is, they do not fall on a periodic (or repeating) pattern. We might say that an ideal glass is a solid, which exhibits an infinite unit cell. The crystallization rate of an undercooled liquid is then specified by the rate of crystal nucleation and by the speed with which the crystal-liquid interface advances. Both of these rates are strongly dependent upon the reduced temperature, T_r and the degree of undercooling ΔT_r which are defined as $T_r = T / T_m$ and $\Delta T_r = (T - T_m)/T_m$ where T_m and T are the equilibrium crystallization and actual absolute temperature, respectively. The rise of similarly construed reduced glass transition temperature T_{rg} is supposed to enhance glass-forming tendency and liquids with this glass temperatures as high as 2/3 T_m, would practically freeze-in within a limited temperature range, certainly if seed-free. Thus they could be undercooled to the glass state in contrast to the liquids with glass transition temperature $T_m/2$ could which could be chilled to the glass state only in relatively small volumes and at high cooling rates. Thus, whether or not the crystallization of an undercooled liquid is bypassed will depend principally upon a set of factors which can be controlled to some extent in the laboratory. Namely by selecting the cooling rate, the liquid volume, the seed density, and set of materials constants, namely, the reduced liquid-solid interfacial tension, the fraction of acceptor sites in the crystal surface and the viscosity-reduced temperature relation which often can be characterized roughly by the reduced glass temperature. The glass-forming tendency will be greater the larger are the cooling rate or enthalpy change and smaller are interfacial tension, etc. Various proposals have been made for correlating glass temperatures. These include scaling of the cohesive energy or the Debye temperature. Kauzman noted that T_{rg}, is approximately 2/3 for a number of simple molecular substances which easily form glasses in bulk. However, another analysis indicates that T_{rg}, must lie considerably below 2/3 for those substances, which do not form glasses in common experience including certain binary chalcogenide systems. Hrubý was a technologist who synthesized and analyzed thousands of chalcogenide compounds and glasses and investigated their thermal behaviors. He proposed the evaluation of glass-forming tendency by means of DTA and triggered development of various other criteria. Besides giving a tribute to the masterwork of a range of scientists, the purpose of this chapter is to revise the meaning and usefulness of various glass forming measures in comparison with the other comparable coefficients and uncover some generalized correlations between glass-forming ability and glass stability

2. GLASS FORMING ABILITY AND GLASS STABILITY

Parameters predicting *glass forming ability* (GFA) and consequent *glass stability* (GS) of its constrained state (of freeze-in melt into glassy state) [1-5] are of a substantial meaning to all those interested in various applications to which glassy materials lend themselves. A time-temperature-transformation (TTT) diagram for material annealing can provide all necessary data, but such data are rarely available and furthermore are often predicted on the assumption of homogeneous nucleation, which is a rather unlikely event in practice. When a glassy matter turn out to be experimentally accessible upon a suitable melt quenching (*critical cooling rate*, R_c) from its *melting point* (T_m) through the *glass transition region* (GTR, defined by the mean glass transition temperature, T_g) certain data became accessible for such a material identification [4-10]. A liquid with good GFA exhibits a low value of R_c for the glass formation, which has remained a long-standing question as to why one liquid exhibits better GFA than another and how we can portray it.

According to the standard nucleation theory, a liquid with a high viscosity between T_g and T_m typically exhibits a high GFA with a low R_c. Since the viscosity at T_g is constant ($\cong 10^{12}$ Pa s), it was postulated that a high value of the *reduced glass-transition temperature*

$$T_{rg} = T_g / T_m \tag{1}$$

would result in a high viscosity in the undercooled liquid state [11, 12], and, consequently, lead to a low R_c. Significantly before the development of any generalized nucleation theory for condensed systems, Tammann [11] called attention to a tendency revealing that the higher the melt viscosity at the melting temperature (T_m), the lower its crystallizability. Qualitatively, this tendency can be explained by an increased inhibition of motion or molecular rearrangement of the basic units of any melt with increasing viscosity. Mentioned by Kauzman [11] and particularly stated by Turnbull, who early indicated that when T_{rg} is larger than 2/3, the homogenous crystal nucleation will be essentially suppressed due to the sluggishness of the crystallization. Therefore, T_{rg} became the earliest criterion to evaluate GFA of a liquid. However, experimental observations indicated that T_{rg} may fail to truthfully predict GFA in many cases, often documented by conventional relationships between R_c and T_{gr} for different sort of glasses, where some good glass formers show a low T_{rg} and vice versa. Referring to a large set of available experimental data obtained for nucleation of several silicate glasses, The reduced glass transition temperature was examined in more details [13] showing that glasses having T_{rg} higher than ~ 0.58 - 0.60 display only surface (mostly heterogeneous) crystallization, while glasses showing volume (homogeneous) nucleation have $T_{rg} < 0.58$ - 0.60. Theoretical values of reduced temperatures were approached by Angell [14] based on extension and extrapolation by means of application of the *Vogel-Fulcher-Tamman* (VFT) viscosity equation. Šesták [10] dealt with a larger extension of reduced quantities and such a difference between liquid and crystalline heat capacities ΔC_{pr}, and evaluated approximate changes of entropy ΔS_r, enthalpy ΔH_r and chemical potential $\Delta \mu_r$ assuming the values T_{or} and T_{rg}. (T_{or} is the reduced Kauzman temperature. To unify the notation, which varies in the papers cited above, in the following text we will use T_x to denote onset crystallization temperature and T_c to denote peak crystallization temperature, as it is common today.

GFA is noticeably related to the ease of the reverse process of devitrification possibly evaluable on basis of the difference between re-crystallization and glass transition temperatures [1, 2]. Therefore, another parameter can be included, which is comfortably monitored by DTA/DSC upon glass heating and which (beside T_g) provides also the nonequilibrium *glass crystallization temperature* laying customary below that for an analogous melt. Than the interval between T_m and T_c i.e., T_x is inversely proportional to GFA and that one between the onset of crystallization T_x and T_g is directly eligible to display GFA. Various examples revealed that this difference varies repeatedly with composition and reaches its maximum value in a composition range which appears to provide "best" glasses.

Even a more sensitive interrelation to the glass formation peculiarities can be found on a basis of the widespread *Hrubý parameter K_H*, developed mostly for chalcogenide systems [15], which are typically available only upon physical preparation of a given type of glass. The Hrubý parameter can be calculated in the following way:

$$K_H = \left(T_x - T_g\right)/\left(T_m - T_x\right) \tag{2}$$

where T_x is the onset of crystallization temperature. This criterion [2] has an almost alike implication as the difference $\left(T_x - T_g\right)$ alone varying, however, more rapidly when crystallization peak is shifted. It takes into account melting temperature, which may not be too significant advantage as T_g and T_m are usually correlated. In order to make the criterion more sensitive, Saad and Poulain [16] took into their consideration the width of a DTA/DSC peak, i.e., the difference between the onset of crystallization T_x and its maximal value T_c accounting, among other features that more stable glasses exhibit broader isotherms and thus a larger crystallization time and therefore a smaller crystallization rate, however, complicated by rather strong dependence of T_c on the heating rate applied. The new criterion appeared $(T_c - T_x)(T_x - T_g)/T_g$ having the unit of [K], but able to be encoded dimensionless if weighted by squared T_g.

The other most popular GS parameters became those that involve major three characteristic temperatures such as the GS factor proposed by Weinberg [17] $K_W = \left(T_c - T_g\right)/T_m$ and more recently, the one envisaged by Lu and Liu [18] as $K_{LL} = T_c/\left(T_g + T_m\right)$. Also, there are other more complex

GS parameters, such as that of Duan et al. [19], which relates kinetics and thermodynamics in the form $K_D(T) = v \exp(-DE/RT)$. In this relation R is the universal gas constant, D is expressed via characteristic temperatures as $D = T_x(T_c - T_x)/\{T_m(T_m - T_g)\}$, v is a frequency factor and E is activation energy for crystallization. However, the values of D, v and E are derived from DSC/DTA curves, which are obtained at different heating rates, and are not fully compatible, so that not acknowledged herewith. Ota et al. [20] related the viscosity at the melting point $\eta(T_m)$, with critical cooling rates, which thus possible to consider $\eta(T_m)$ as a GS parameter, as originally proposed by Tamman [11] many years ago.

In the recent years, various parameters have been investigated, such as those in the papers [21-31]. These parameters are related to the three characteristic temperatures (T_g, T_c i.e., T_x and T_m, i.e., liquidus temperature-T_l). The authors of all of these papers assume that the parameters present reliable GFA criteria. The possible correlation between GS and GFA is of the great importance. Hence, this topic was the subject of theoretical work as well as different experiments. Weinberg in [17] used standard theoretical expressions for crystal nucleation and growth rates as well as the *Johnson-Mehl-Avrami-Yerofeev-Kolmogorov* (JMAYK) [2, 9] theory and considered homogeneous nucleation and screw dislocation growth in stoichiometric glasses. In this paper, the trends in GFA and GS were compared with systematic changes in the melting entropy, ΔS_m, and the viscosity parameters in the *Vogel–Fulcher–Tamman* (VFT) expression [32-35] in terms of the reduced temperature $T_r = T/T_m$ and $\eta = \eta_o \exp[b/(T_r - T_{or})]$, using the concept of the reduced Kauzman temperature, T_{or} and the values of the pre-exponential, η_o as well as the constant b. The result was a conclusion that GFA and GS, defined by $(T_c - T_g)/T_m$ are poorly related.

Using again the JMAYK theory, Weinberg and a subsequent work [22], derived the time necessary to crystallize a minimum detectable fraction. This development was again based on classical homogeneous nucleation and screw dislocation growth in stoichiometric glasses. Time criteria were used to assess GFA and test the reliability of two particular GS parameters, given by the previously mentioned expressions:

$$\left(T_x - T_g\right)/T_m \text{ and/or } \left(T_x - T_g\right)\left(T_c - T_x\right)/T_m. \qquad (3)$$

Weinberg [17] calculated these data and observed that the stability of glasses having parallel viscosity curves $\eta(T)$ could be qualitatively evaluated by the two above expressions, but they are not quantitatively reliable and none of them appears to be superior. In spite of it, the paper also shows that for glasses for which $\eta(T)$ significantly differs in the region of T_g the stability criteria can be delusive.

3. EXPRESSING THE HRUBÝ PARAMETER USING TEMPERATURE RELATIONS: $R = T_c/T_G$ AND $M = T_M/T_G$

After Weinberg [17] had published his paper, Cabral et al. [21] used experimental values and found a correlation between the Hrubý parameter of glass stability (GS) and glass forming ability (GFA). However, the results of Cabral were contradictory in relation to the theoretical calculations of Weinberg. Avramov et al. [22] decided to check these two approaches. They extended the calculations of Weinberg [17] but tested a different assumption, which is supported by experimental data. In the paper [22] by Avramov et al. demonstrated that GFA and GS follow the same trend and are directly related. Recently, several important studies have appeared, which demonstrate correlation between the GS parameters and critical cooling rate R_c, or between the GS parameters and maximum section thickness, i.e., the diameter D_{max}, by which the glass forming ability are estimated. In the papers [23-25], a very good correlation is shown between GFA and GS parameters, which are based on the three characteristic temperatures such as the Hrubý (K_H) [15], Weinberg (K_W) [17], and Lu-Liu (K_{LL}) [18] parameter.

Nascimento et al. [26] proved that for the oxide glasses a very good correlation between the Hrubý parameter and GFA exists. Also, very good correlation between K_{LL} and GFA has been found for the oxide glasses in [25, 26]. The R_c is a quantity hard to measure and the D_{max} cannot also be measured in a sufficiently precise way. Hence, of essential importance are the correlations between the GS parameters expressed via characteristic temperatures and GFA [24-27]. These temperatures are usually determined by DTA and DSC analyses. A satisfactory degree of correlation between the determined GS parameter and R_c (or D_{max}) would allow one to use the given GS parameter to assess GFA.

In spite of the parameters discussed above, which are calculated using the three characteristic temperatures, the reduced glass transition temperature shows the weaker correlation with GFA. The work of Nascimento [26] proves that T_{rg} has the weaker correlation with GFA for oxide glasses than K_H, K_W and K_{LL}. Also, Lu and Liu showed that T_{rg} has the weaker correlation with GFA than their parameter K_{LL} for glasses analyzed in [23]. For this reason, in the following text we will analyze the Hrubý parameter as well as some other, often used parameters [2, 29], which depend on the three characteristic temperatures for the sake of simplicity disregarding many others [27-31].

Mentioning the work [28], which stresses that GS parameters exploited to estimate GFA as an enough sensitive factor? When switching from one glassy system to another one, the glass resistance in relation to devitrification changes, and the GS parameters change as well. Larger values of the K_H, K_W and K_{LL} parameters imply higher stability of the glass in respect to devitrification. Also, when comparing one glass to another, it is essential to know how large the relative change of the given parameter is and how it can be compared with the relative change of other GS parameters. In other words, it is necessary to know which of the GS parameters shows the fastest change.

Our starting point is the fact that all the parameters, K_H, K_W and K_{LL}, include the three characteristic temperatures [2, 29]. For this reason, we can express them in a somewhat different way – using the ratios of the temperatures.

$$m = T_m/T_g \text{ and } r = T_c/T_g. \tag{4}$$

It is necessary to point out that the following relations always hold: $m > 1$, $r \geq 1$, $m > r$.

In doing so, we assume that in defining both K_H and K_{LL} it is possible to replace onset crystallization temperature T_x with maximum crystallization peak temperature T_c as was shown in the paper of Nascimento etal [26].

After a simple mathematical transformations [29] one obtains

$$K_H = \frac{r-1}{m-r} \quad K_W = \frac{r-1}{m} \quad K_{LL} = \frac{r}{m+1} \tag{5}$$

In this way, by using the substitutions r and m, the GS parameters are expressed indirectly via the reduced glass transition temperature T_{rg} and super cooled region ΔT_{xg}. It is because the parameter m represents the reciprocal value of T_{rg}, and the parameter r can be correlated to ΔT_{xg}, as was shown in the work of Lu and Liu [23] and others [30, 31].

Namely, in order to enable the comparison for different glasses, the value of the super cooled region is divided by T_g [23], which gives out $(T_x - T_g)/T_g = r - 1$. Mondal etal [30] used the same normalization, too, but proposed that T_x/T_g (in our cases r) can also be considered as a measure of the thermal stability of glass. Further the factor of crystallization resistance $T_g/(2T_x - T_g)$ was introduced which can be expressed as $1/(2r-1)$. From a mathematical point of view, the introduction of the ratios r and m allow the reduction of the number of independent variables by which the parameters K_H, K_W and K_{LL} are defined. As an alternative of the three characteristic temperatures, we now use their two ratios (r and m) as independent variables.

4. RELATIVE CHANGES OF GS PARAMETERS

In order to derive expressions for the relative changes of these parameters, as shown in our papers [2, 29], we first took logarithms of expressions for K_H, K_W and K_{LL} and then differentiated the obtained values

$$\frac{dK_H}{K_H} = \frac{dr}{r-1} - \frac{dm}{m-r} + \frac{dr}{m-r} \qquad \frac{dK_W}{K_W} = \frac{dr}{r-1} - \frac{dm}{m} \qquad \frac{dK_{LL}}{K_{LL}} = \frac{dr}{r} - \frac{dm}{m+1} \qquad (6)$$

By comparing the right hand side expressions in Eq. (6), as given in [29], it holds that if the condition $dr > dm$ is fulfilled, that is if $d(T_c/T_g) > d(T_m/T_g)$ is satisfied, then it will always hold that

$$\frac{dK_H}{K_H} > \frac{dK_W}{K_W} > \frac{dK_{LL}}{K_{LL}} \qquad (7)$$

Such a relation of the relative changes of the GS parameters will also hold when a less stringent condition is satisfied, i.e., when $dr/r > dm/m$. Hence, we can say that if the relative change of the ratio of the crystallization temperature and glass transition temperature is greater than the relative change of the reciprocal value of the reduced glass transition temperature, then the change of the Hrubý parameter will be the greatest, followed by the change of the Weinberg parameter, while the change of the Lu-Liu parameter will be the smallest.

If the condition $dr/r > dm/m$ is not fulfilled, then the following relation of the relative changes of the GS parameters will hold: $dK_W/K_W > dK_H/K_H > dK_{LL}/K_{LL}$. As can be seen from this relation and Eq. (7) the relative change of the Lu-Liu parameter will always have the smallest value, and will never be greater than the relative change of the Hrubý parameter.

Table 1. The values of the m and r, as well as the values of characteristic terms in calculation of the maximal relative change of the Hrubý parameter for oxide glasses [29]

glass	m	r	Δr	Δm	$\Delta r/(r-1)$	$\Delta m/(m-r)$	$\Delta r/(m-r)$	$(\Delta K_H/K_H)_{max}$
G	1,69062	1,44579	0	0	0	0	0	0
NS$_2$	1,59084	1,29681	0,148988	0,09977	0,50196	0,33932	0,50670	1,347993
CMS$_2$	1,68250	1,22244	0,223351	0,00811	1,00406	0,01763	0,48548	1,507182
PS	1,53402	1,34615	0,099644	0,15659	0,28786	0,83354	0,53039	1,651792
M$_2$A$_2$S$_5$	1,25581	1,18418	0,261611	0,43480	1,42036	6,07035	3,65238	11,14310
CAS$_2$	1,62188	1,17704	0,268751	0,06873	1,51797	0,15451	0,60415	2,276645
LS$_2$	1,76010	1,24528	0,200514	0,06948	0,81748	0,13497	0,38948	1,341938
LB$_2$	1,56191	1,07057	0,375225	0,12870	5,31687	0,26194	0,76367	6,342485

Where GeO$_2$ (G); Na$_2$O $\cdot$ 2SiO$_2$ (NS$_2$); CaMg $\cdot$ 2SiO$_2$ (CMS$_2$); Pb $\cdot$ SiO$_2$ (PS); 2MgO $\cdot$ 2Al$_2$O$_3$ $\cdot$ 5SiO$_2$ (M$_2$A$_2$S$_5$); CaO 2Al$_2$O$_3$ $\cdot$ 2SiO$_2$ (CAS$_2$); Li$_2$O $\cdot$ 2SiO$_2$ (LS$_2$); Li$_2$O $\cdot$ 2Ba$_2$O$_3$ (LB$_2$).

The above theoretical derivations were tested on two series of oxide glasses and one series of chalcogenide glasses [29]. Results of testing are completely in agreement with the theoretical results.

The next step was to establish a relation between the maximal value of relative changes of the Hrubý (H), Weinberg (W), and Lu-Liu (LL) parameters. In order to obtain the maximal value of relative changes, minus signs in expressions is presented, in Eq. (6), to be replaced by the plus ones. Thus, we obtain

$$\left(\frac{dK_H}{K_H}\right)_{\max} = \frac{dr}{r-1} + \frac{dm}{m-r} + \frac{dr}{m-r} \quad \left(\frac{dK_W}{K_W}\right)_{\max} = \frac{dr}{r-1} + \frac{dm}{m} \quad \left(\frac{dK_{LL}}{K_{LL}}\right)_{\max} = \frac{dr}{r} + \frac{dm}{m+1} \tag{8}$$

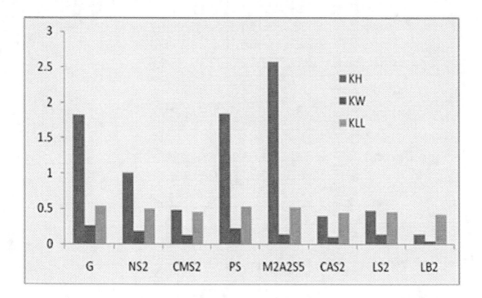

Figure 1. Histograms of the K_H, K_W and K_{LL} parameters for oxide glasses from Table 1

After comparing right hand side expressions in Eq. (8) the following relation between the maximal values of relative changes of the parameters considered above will hold:

$$(\frac{dK_H}{K_H})_{\max} > (\frac{dK_W}{K_W})_{\max} > (\frac{dK_{LL}}{K_{LL}})_{\max} \qquad (9)$$

The maximal relative change of the Hrubý parameter will be the greatest, followed by the change of the Weinberg parameter, while the change of the Lu-Liu parameter will be the smallest, see Figure 1.

5. MODIFIED ANGELL PLOT OF VISCOUS FLOW AND FRAGILITY CONCEPT

Tamman [11] already thought that the fragility of a supercooled liquid can quantify the extent to which the viscosity of the liquid approaches an Arrhenius-like temperature dependence. According to Angell [32, 35-37], the kinetic fragility parameter m was defined as

$$m = \frac{d(\log \eta)}{d(T_g / T)}\bigg|_{T=T_g} \qquad (10)$$

In the original Angell plot the logarithm of viscosity $\log \eta$ is plotted as a function T_g/T and m is the slope of viscosity curve near glass transition temperature. The large m value in the Angell plot indicates that liquid behaves fragile

It is well known feature of the Vogel-Fulcher-Tammann (VFT) functions [11, 39-41] in the Angell plot, but, it is possible a different approach to fragility. We proposed [33] the plot of $\log \eta_{VFT}$ in relation to $t = T_g/(T-T_o)$, where T_o is temperature at which the configurational entropy becomes zero. This temperature is equivalent to either the Kauzmann temperature T_K [4, 7] or a finite temperature T_o of the VFT equation. The plot of $\log \eta_{VFT}$ in relation to $t = T_g/(T-T_o)$ is a group of linear functions. The parameter m_1 defined as $m_1 = d(\log \eta)/d(T_g/(T-T_o))\big|_{T=T_g}$ was determined as a slope of the line. The higher the value of m_1, the "stronger" the glasses. The relationship between the parameter m in the Angell plot and the parameter m_1 in the T_g scaled VFT plot is $m_1 = m(1 - T_o / T_g)^2$. With the glasses where the following relationship $m /(m - 17) = T_g / T_o$ is valid, it also holds that $m_1 = 289/ m$

Some different approaches concerning fragility concept presented in the work of Schmelzer et al. [43, 44]. The authors do not connect the fragility with glass transition but relate it for a specific temperature of maximum rates of nucleation, growth, and overall crystallization in dependence on the particular process analyzed

On the other hand Laughlin and Uhlmann [45] classified the presentation of viscosity in terms of T_m/T as a kind of "superior normalization" of viscosity data but went over then to T_m/T considering it as "difficult to rationalize a priori the dependence of liquid on characteristics other than those of the liquid phase alone".

6. REDUCED GLASS TRANSITION TEMPERATURE AND FRAGILITY AS PARAMETERS FOR ESTIMATING GLASS FORMING ABILITY

In addition to the already described K_H, K_W and K_{LL} parameters, recently a number of new GFA parameters defined by characteristic temperatures have been proposed [37-45]. However, we assumed [2, 46] that a correlation between GFA and the liquidus viscosity η_L exists. For this reason, the relationship between characteristic temperatures that appears in η_L can define the GFA parameter.

In the Vogel-Fulcher-Tammann (VFT) equation [11, 39-41], the dependence between viscosity and absolute temperature is expressed as

$$\log \eta = \log \eta_\infty + \frac{B}{T - T_o} \cdot \frac{1}{\ln 10} \ . \tag{11}$$

In this equation B is constant, η_∞ is pre-exponential factor and T_o is temperature at which the configurational entropy becomes zero. This relation keeps association with kinetic studies detailed in [46] which is not a subject of our communication

We transformed this equation for liquids viscosity [42] in

$$\log \eta_L = \log \eta_\infty + 16 \frac{(T_g - T_o)}{(T_l - T_o)} \tag{12}$$

Because GFA is proportional to liquids viscosity η_L, it follows that

$$GFA \approx \log \eta_\infty + 16 \frac{(T_g - T_o)}{(T_l - T_o)} \tag{13}$$

The F_K parameter for estimating of GFA can be defined [47] as

$$F_K = \frac{(T_g - T_o)}{(T_l - T_o)} \tag{14}$$

It should be mentioned that this parameter appears in the work of Takeuchi et al. [48], but without any particularization. However, in our development, there is no indication that F_K relates to T_{rg} so that we can substitute reduced glass transition temperature as the GFA parameter [42, 48]. For this reason we also defined [33, 47] a new parameter F_{KA} as the sum of F_K and T_{rg} as

$$F_{KA} = \frac{16}{17} F_K + \frac{1}{17} T_{rg} \tag{15}$$

Bohmer and Angell [49] obtained a relation $m/(m-16) = T_g/T_o$ from which follows that $m = 16 T_g /(T_g - T_o)$. The factor "16" that appears in this relationship has a deep physical meaning. This is the minimal value of kinetic fragility parameter m. Expressing T_o from this relation and substituting it in eq. (14) for F_K, we obtain

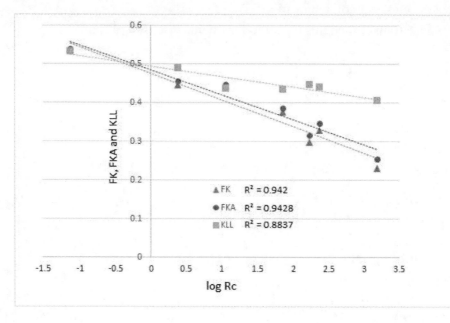

Figure 2. Glass forming ability parameters F_K, F_{KA} as well as parameter K_{LL} as a function of log R_c for the glass forming systems from [33, 47].

$$F_K = \frac{16}{\dfrac{m}{T_{rg}} - m + 16}$$ (16)

There are some further papers dealing with glass-formation data due to a notice [50-61], however, rather extensive to include in more details. Let us continue by comparing the correlations of F_K and F_{KA} with the correlation of GS parameter K_{LL} which is proposed by Lu and Liu, with critical cooling rate. The linear correlation of F_K, F_{KA} and K_{LL} with log R_c, as well as the corresponding values of R^2 factors for selected silicate glasses, are presented in Figure 2. It is clear that K_{LL} has lower correlation with R_c ($R^2 = 0.8837$) than F_K and F_{KA}.

When a reduced glass transition temperature T_{rg} becomes constant, F_K and F_{KA} are single functions of fragility and behave as $1/m$. These parameters increase with the increase of GFA, for this reason glasses with the same reduced glass temperature will have higher glass forming ability if they have the lower fragility. But the question that arises is what are the factors governing fragility

7. STRETCHING EXPONENT AGAINST CONFIGURATIONAL ENTROPY: WHICH FACTOR HAS GREATER CONTRIBUTION TO KINETIC LIQUID FRAGILITY?

Gupta et al. [62] developed the following expression for fragility parameter m:

$$m = m_o\left(1 + \frac{\partial \ln S_c(T)}{\partial \ln T}\bigg|_{T=Tg}\right) + \frac{1}{\ln 10}\left(\frac{2\pi^2}{12(\beta_{Tg})^2 + \pi^2[1 - (\beta_{Tg})^2]}\right)\frac{\partial \ln \beta(T)}{\partial \ln T}\bigg|_{T=Tg} \quad (17)$$

Where $S_c(T)$ is the configurational entropy, T_g is the glass transition temperature and β is the stretching exponent governing the non-exponentiality of the relaxation process [63-73].

Eq (17) shows that two terms exist in fragility of a supercooled liquid. The first term is thermodynamic depending on the change of configurational entropy, while the second one is kinetic, and depends upon non-exponentiality of the relaxation function.

We begin the analysis of the kinetic term in our previous work [74], expressing the derivative of stetching exponent as $\frac{\partial \ln \beta(T)}{\partial \ln T} = \frac{T}{\beta(T)}\frac{\partial \beta(T)}{\partial T}$.

From the results obtained by Palato et al. [75], data from the work Cangialosi et al. [76] and our results presented in [74], it follows that in the interval $1 \leq T/T_g < 1.1$ for given m, we can take as approximation that the stretching exponent increases as a linear function of T/T_g. As shown in [64] assuming that $\beta(T)$ is a linear function of T/T_g, that is, $\beta(T) = a\dfrac{T}{T_g} + b$ with coefficients $a > 0$ and $b > 0$, the kinetic term on fragility can be expressed as

$$m_K = \frac{1}{\ln 10}\left(\frac{2\pi^2}{12(\beta_{Tg})^2 + \pi^2[1 - (\beta_{Tg})^2]}\right)\frac{a}{a+b} \qquad (18)$$

that is $m_K = f(\beta_{Tg})\dfrac{a}{a+b}$. The values of function $f(\beta_{Tg})$, as was presented in [74], will be in the interval from 0.714 to 0.869. The ratio $a/(a + b)$ is always smaller than 1 because a and b are positive, for this reason, the kinetic term m_K is always smaller than 1.

Thermodynamic term in Eq. (17) is

$$m(S_c) = m_o\left(1 + \frac{\partial \ln S_c(T)}{\partial \ln T}\bigg|_{T=Tg}\right) \qquad (19)$$

The change of configurational entropy, can be expressed as [62]: $\dfrac{\partial \ln S_c(T)}{\partial \ln T}\bigg|_{T=Tg} = \dfrac{C_p^{conf}(T_g)}{S_c(T_g)}$, where $C_p^{conf}(T_g)$ is the configurational heat capacity of the liquid at the glass transition temperature. As result, for thermodynamic term it holds

$$m(S_c) = m_o\left(1 + \frac{C_p^{conf}(T_g)}{S_c(T_g)}\right). \qquad (20)$$

The thermodynamic term $m(S_c)$ is not smaller than m_o because the values of $C_p^{conf}(T_g)$ and $S_c(T_g)$ cannot be negative. The lowest limit of $m(S_c)$ is m_o, what is the case when configurational entropy is constant, that is

$\dfrac{\partial \ln S_c(T)}{\partial \ln T}\bigg|_{T=Tg} = 0$. The value of parameter m_o that corresponds to the strongest liquid can be calculated by relaxation time τ as $m_o = \log \tau_g/\tau_o$ [77], where $\tau_g = 100$s and $\tau_o \approx 10^{-14}$s. This means that $m_o \approx 16$ and for this reason the value of the thermodynamic term is $m(S_c) \geq 16$. Since the kinetic term m_K is always smaller than 1 it is obvious that the thermodynamic term plays the main role in fragility and kinetic term can be neglected. Therefore, with good approximation it is possible to express the fragility index m as $m \approx m(S_c)$,

That is

$$m = m_o\left(1+\frac{\partial \ln S_c(T)}{\partial \ln T}\bigg|_{T=Tg}\right) = m_o\left(1+\frac{C_p^{conf}(T_g)}{S_c(T_g)}\right) \qquad (21)$$

8. THE FUNCTION OF TEMPERATURE DEPENDENCE OF CONFIGURATIONAL ENTROPY

However, a realistic determination of the configurational entropy and configurational heat capacity as a function of temperature is not quite resolved and is still a subject of discussion [78-84]. The question is how to determine the function of temperature dependence of configurational entropy $S_c(T)$, the derivative of which appears in Eq. (21). One possible way to resolve this problem is to obtain the $S_c(T)$ function from the equations of viscous flow. Different models have been suggested for describing the temperature dependence of viscosity. The most popular viscosity models describing glass forming liquids are the Adam–Gibbs (AG) equation [85], the Vogel-Fulcher-Tammann (VFT) equation [11, 39-41], the Avramov and Milchev model (AM) [86-88] and the recently proposed MYEGA equation by Mauro etal [89]. These models have been compared in several studies [90-93]. The common approach in these studies was to assume that there are differences between the functions of viscous flow and consequently in the values of log η (T) calculated on the basis of each model. In most cases, the authors rewrote the equations of viscosity from different models to two identical quantities: glass transition temperature and fragility, and compared them. Contrary to this, Sipp et al. [94] derived the expression for $S_c(T)$ assuming that the AG equation and the VFT equation are equivalent. Recently, Yue [81] derived the $S_{cVFT}(T)$ function from the VFT equation using a concept of iso-structural viscosity. In this paper, we will analyze and compare these two expressions for temperature dependence of configurational entropy and their influence on the thermodynamic term [58, 93-98].

Analyzing high viscosity data for 3D network, Sipp et al. [94] obtained the expressions for the temperature dependence of the configurational entropy and configurational heat capacity. The authors assumed that viscosity should be the same regardless of whether it is expressed by the AG equation or by the VFT equation, that is $\ln \eta(T)_{VFT} = \ln \eta(T)_{AG}$.

The VFT and AG functions are respectively $\ln \eta(T)_{VFT} = A_{VFT} + B_{VFT}/(T-T_o)$ and $\ln \eta(T)_{AG} = A_{AG} + B_{AG}/(TS_c(T))$, where $A_{VFT}, B_{VFT}, A_{AG}, B_{AG}$ and T_o are adjustable constants. Based on that at glass transition temperature, the equality $\ln \eta(T_g)_{VFT} = \ln \eta(T_g)_{AG}$ is also valid and additionally assuming that pre-exponential parameters A_{AG} and A_{VFT} are equal. Sipp et al. [94] obtained the following equation for the configurational entropy

$$S_c(T) = S_c(T_g)\frac{1-T_o/T}{1-T_o/T_g} \qquad (22)$$

and the following equation for the configurational heat capacity

$$C_p^{conf}(T) = S_c(T_g)\frac{T_o/T}{1-T_o/T_g} \qquad (23)$$

For $T = T_g$, Eq. (23) can be rewritten as

$$C_p^{conf}(T_g) = S_c(T_g) \frac{T_o/T_g}{1 - T_o/T_g} \tag{24}$$

Substituting this expression in Eq. (21), for the fragility index we obtain

$$m = m_o\left(1 + \frac{T_o/T_g}{1 - T_o/T_g}\right) \tag{25}$$

That is

$$m = m_o\left(1 + \frac{T_o}{T_g - T_o}\right) \tag{26}$$

Eq. (26) is equivalent to the equation for fragility m, derived in another way in the study by Smedskajer et al. [95].

Using a simple transformation from Eq. (26), we can obtain the expression proposed by Bohmer et al. [38] that $m = 16T_g/(T_g - T_o)$ in which the minimum value of fragility is $m_o = 16$. This expression allows us to calculate the values of m in a simple way. In the study by Sipp et al. [94], the configurational entropy $S_c(T)$ is the same for both the AG and the VFT model. The temperature dependence of configurational entropy for a given alloy is a universal function and independent on these two models.

Applying a different approach Yue [81] considers the temperature dependence of the equilibrium viscosity for basalt liquid and window glass liquid. The author shows that the differences between the AG and the VFT functions of viscous flow are rather small. Nevertheless, Yue has not assumed that these two log η functions are equal. Instead, the VFT function was transformed in an AG fashion. The calculated temperature dependences of the configurational entropy in have different plots, depending on the model which is used. From the transformed VFT function, it holds that the temperature dependence of the configurational entropy $S_{cVFT}(T)$ is

$$S_{cVFT}(T) = \frac{B_{AG}}{B_{VFT}}\left(1 - \frac{T_o}{T}\right) \tag{27}$$

that is

$$S_{cVFT}(T) = S_{c\infty Y}\left(1 - \frac{T_o}{T}\right) \tag{28}$$

where $S_{c\infty Y} = B_{AG}/B_{VFT}$.

Analogously, Eq. (22) for $S_c(T)$ it is obtained [94] but rewritten as

$$S_c(T) = S_{c\infty S}\left(1 - \frac{T_o}{T}\right) \tag{29}$$

where $S_{c\infty S} = S_c(T_g)/(1 - T_o/T_g)$. Both equations (Eq. (28) and (29), show the same temperature dependence of the configurational entropy, that is, the proportionality with $(1-T_o/T)$. The only difference between these two expressions is coefficient multiplication. Both coefficients, $S_{c\infty S}$ and $S_{c\infty Y}$, present the maximum of the configurational entropy at an infinite high temperature. It holds from Eqs. (28) and (29) that configurational entropy $S_c(T)$ and $S_{cVFT}(T)$ will be zero at some finite temperature T_o.

The same temperature dependence in Eq. (28) and Eq. (29) is not surprising for the following reasons. In the fundamental paper [85] Adam and Gibbs assumed that the configurational heat capacity is approximately independent of temperature. According to the AG entropy function, the configurational entropy would vanish at an equilibrium second-order transition temperature T_2 equivalent to the Kauzmann temperature T_K[11] or T_o of the VFT equation. This was discussed in [89] where the authors introduced a new model of viscosity of glass-forming liquids that does not predict the vanishing of configurational entropy at finite temperature. Assuming that C_p^{conf} is constant and $T_K = T_o$, through the integration of C_p^{conf} over the temperature range from T_K to T, one can get the configurational entropy and introduce it into the AG equation, obtaining finally the AG equation with the same form as the VFT equation. However, C_p^{conf} is not necessarily constant with respect to temperature. Yue [81] obtained an equation for configurational entropy which implies that C_p^{conf} is proportional to $1/T$. This equation can be obtained substituting temperature T_K by T_o in the AG model implying again that the AG and the VFT equation are identical if $T_K = T_o$. The proportionality $C_p^{conf} \sim 1/T$ also holds from Eq. (23) which was obtained by Sipp et al. assuming that the AG equation and the VFT are equivalent. To compare expressions for temperature dependence of configurational entropy $S_c(T)$ which was proposed by Sipp et al. [94] and $S_{cVFT}(T)$ obtained by Yue [81] we calculated these functions for GeO_2, SiO_2, $Na_2O \cdot Al_2O_3 \cdot 6SiO_2$ (Albite, Ab), $CaO \cdot Al_2O_3 \cdot 2SiO_2$ (Anorthite, An) and $CaO \cdot Mg_2O \cdot 2SiO_2$ (Diopside, Di). Table 2 contains input data to calculate the values of coefficients $S_{c\infty S}$ and $S_{c\infty Y}$, as well as the function of $S_c(T)$ and $S_{cVFT}(T)$ for the considered glass formers.

Table 2. Values of coefficients $S_{c\infty S}$ and $S_{c\infty Y}$ for glasses whose characteristic temperatures, configurational entropies and configurational heat capacities at T_g, were taken from reference [94]

Glass	T_g (K)	T_o (K)	$10^{-5} B_{AG}$ J/(oxmol)	$10^{-4} B_{VFT}$ (K)	$S_c(T_g)$ J/(oxmol·K)	$C_p(T_g)$ J/(oxmol·K)	$S_{c\infty S}$	$S_{c\infty Y}$
GeO2	816	199	1.929	2.241	6.5	2.1	8.596	8.608
SiO2	1452	529	2.863	3.353	5.4	3.1	8.495	8.539
Ab	1084	383	3.230	2.778	7.5	4.1	11.598	11.627
An	1129	768	4.222	1.428	9.7	20.4	30.336	29.566
Di	994	716	2.876	1.065	7.6	19.5	27.174	27.005

Figure 3. exemplifies the plot of $S_{cVFT}(T)$ and $S_c(T)$ versus T obtained from Eq. (28) and Eq. (29), respectively. The dependence of ln Sc(T) on ln (T) for GeO_2, SiO_2, Ab, An and Di is presented in Figure 4. From Table 2, it is evident that the values of the coefficients $S_{c\infty S}$ and $S_{c\infty Y}$ are almost identical for each glass former, with negligible difference between them. Thus, it is expectable that the plots of $S_c(T)$ and $S_{cVFT}(T)$ functions will almost overlap for all the considered alloys. Figure 3, which presents the

plots of $S_c(T)$ and S_{cVFT} (T), confirms this. Therefore, we will get the same value of the fragility index using either $S_c(T)$ or S_{cVFT} (T) in Eq. (21). For this reason, only the plot of ln Sc(T) vs. ln (T) has been presented in Figure 4.

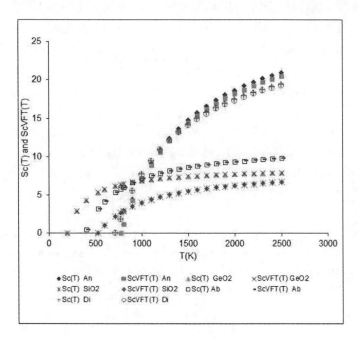

Figure 3. Temperature dependence of configurational entropy $S_{cVFT}(T)$ and $S_c(T)$ obtained from Eqs. (28) and (29) respectively, for the glass formers from [94, 95].

As shown in Figure 4, the dependence of ln Sc(T) on ln (T) for An and Di increases much faster than for GeO_2, SiO_2 and Ab. Therefore, the values of $\partial \ln Sc(T)/\partial \ln T \mid_{T=Tg}$ for An and Di will be higher than for GeO_2, SiO_2 and Ab. These two values will be similar because curves for An and Di almost overlap around glass transition temperatures (ln T_g is 7.056 and 7.00 for An and Di respectively). Consequently, using Eq. (21), the fragility index m for An and Di will be similar and, at the same time, higher than for the rest of glass formers from Figure 3.

From Figure 4 it is obvious that the curves for GeO_2, SiO_2 and Ab increase very slowly with the increasing values of ln (T). This is most significant for GeO_2. Therefore, the values of m for these alloys will be only slightly higher than m_o.

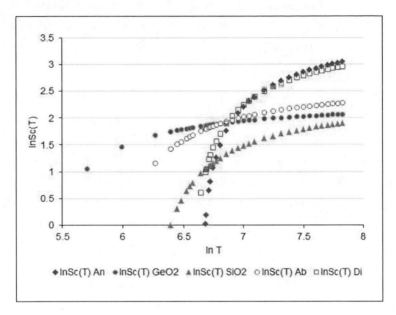

Figure 4. The plot of $\ln S_c(T)$ vs. $\ln(T)$, for the glass formers from [94].

The values of fragility index m obtained from Eq. (26) and data from Table 1 for GeO_2, SiO_2, Ab, An and Di are 21.2, 25.2, 24.7, 50.03 and 57.2 respectively. It can be concluded that from the dependence of $\ln S_c(T)$ vs. $\ln T$ it is possible to successfully predict the relations between the values of m for different glass formers.

CONCLUSION

Despite not clearly and specifically definable state on a previously experimentally 'get ready' glass the investigational glass transition temperature T_g is only of a fundamental usefulness. We found that the relations between the magnitude of the change of the ratios of crystallization and melting temperature with glass transition temperature, that is T_c/T_g and T_m/T_g, determine the order of the values of relative change of GS parameters

dK_H/K_H, dK_W/K_W and dK_{LL}/K_{LL}. Moreover, it takes place in the Angell plot when log η is plotted as a function T_g/T. We proposed a innovative plot of log η_{VFT} in relation to $t = T_g/(T-T_o)$ which subsists a group of linear functions where the slope of the line is the parameter m_1 defined as $m_1 = d(\log \eta)/d(T_g/(T-T_o))\big|_{T=Tg}$. The higher the value of m_1, the "stronger" the glasses occur [96].

We defined new GS parameters F_K and F_{KA} which include fragility and reduced glass transition temperature. The linear correlation of F_K and F_{KA} with log R_c is better than a correlation of K_{LL} [2, 29, 47].

Assuming that stretching exponent governing the non-exponentiality of relaxation process is a linear function of T/T_g we obtained that the contribution of the kinetic term in fragility can be neglected [33,47]. The thermodynamic term, which has a dominant role in fragility, can be determined by the expressions proposed by Sipp et al. [94] for configurational heat capacity and configurational entropy which was analyzed elsewhere [97, 98]. The glass transition temperature appears as a notable parameter in temperature dependence of configurational entropy $S_c(T)$. From the dependence of ln $S_c(T)$ vs. lnT it is possible to successfully predict the relations between the values of m regarding the status of different glass formers [33].

Detailed analysis of various performance of diffusion process can be found elsewhere [86, 90, 99]. Detailed examination of temperature determination using popular experimental methodic DTA throughout this chapter is systematically described in our book chapter [100].

ACKNOWLEDGMENTS

The present work was supported the CENTEM project No. CZ.1.05/2.1.00/03.0088 that is co-funded from the ERDF as a part of the MEYS - Ministry of Education, Youth and Sports OP RDI Program in the follow-up sustainability stage supported through the CENTEM PLUS LO 1402 as well as the internal university project 2019 sponsored by the West

Bohemian University in Pilzen which is due to the recognition. Thanks are due to the Serbian Ministry of Science and Technological Development when subsidizing the projects No. III 45021 and No.172059 carried out by University of Novi Sad. Previous cooperation and friendship of Prof. Jaroslav Šesták (1938 -) with Dr. Arnošt Hrubý (1919 - 2014) is greatly cherished and treasured.

REFERENCES

[1] Queiroz, C. A. and Šesták, J. (2010). Aspects of the non-crystalline state. *Phys Chem Glass Eur J Glass Sci Technol B*, *51*, 165–168, and Šesták, J., Queiroz, C., Mareš, J. J. and Holeček, M. (2011). Some aspects of vitrification, amorphisation, disordering and the extent of nano-crystallinity in the book *Glassy, amorphous and nanocrystalline materials* (Šesták J, Mareš JJ, Hubík P, eds). Berlin: Springer, pp. 59-75.

[2] Kozmidis-Petrovic, A. and Šesták, J. (2012). Forty years of the Turnbull reduced glass-transition temperature and Hrubý glass-forming coefficient, *Thermal analysis of Micro-, nano- and non-crystalline materials*, Springer Berlin, pp. 7-40, and (2012). Forty years of the Hruby′ glass-forming coefficient via DTA when comparing other criteria in relation to the glass stability and vitrification ability. *J Therm Anal Calorim*, *110*, 997–1004.

[3] Illeková, E. (1993). A Generalized Model of Structural Relaxation in Metallic and Chalcogenide Glasses. *Key Eng. Mater.*, 81-83, 541-548, and (1993). Review of Structural Relaxation Models with the Mutual Correlation of their Activation Enthalpies. *Int. J. Rapid Solidific*, *8*, 195-224.

[4] Kauzmann, W. (1948). The nature of the glassy state and the behavior of liquids at low temperatures. *Chem. Rev.*, *43*, 219–256, and Gibbs, J. H. and DiMarzio, E. A. (1958). Nature of the Glass Transition and the Glassy State. *J. Chem. Phys.*, *28*, 373-383.

[5]   Wunderlich, B. (2007). Glass transition as a key to identifying solid phases. *J Appl Polym Sci.*, *105*, 49–59, and Gutzow, I., Pascova, R. and Schmelzer, J. W. P. (2010). Glass Transition Behavior: A Generic Phenomenological Approach. *Int. J. Appl. Glass Science*, *1*, 221-236.

[6]   Schmelzer, J. W. (2012). Kinetic criteria of glass formation and the pressure dependence of the glass transition temperature. *J. Chem. Phys.* *136*, 074512-1/074512-11, and Priestley, R. D., Cangialosi, D. and Napolitano, S. (2015). On the equivalence between the thermodynamic and dynamic measurements of the glass transition in confined polymers. *J. Non-Cryst. Solids*, *407*, 288–295.

[7]   Černošek, Z., Holubová, J. and Černošková, E. (2005). Kauzmann temperature and the glass transition, *J. Optoelectr. Mater.*, *7*, 2941–2944, and Liška, M., Černošek, Z., Holubová, J., Černošková, E. and Vozár, L. (2010). New features of the glass transition revealed by the StepScan DSC. *J Thermal Anal Calorim*, *101*, 189-194.

[8]   Hutchinson, J. M. (2009). Determination of the glass transition temperature: Methods correlation and structural heterogeneity. *J. Therm. Anal. Calorim.*, *98*, 579–589, and He, J., Liu, W. and Huang, Y. X. (2016) Simultaneous Determination of Glass Transition Temperatures of Several Polymers *Open Biomater Res*, *PLoS ONE*, *11*, e0151454. https://doi.org/10.1371/journal.pone.0151454.

[9]   Svoboda, R., Málek, J. and Šesták, J. (2017). Thermo-kinetic phenomena occurring in glasses: their formalism and mutual relationships. Chapter 11 in book: *Thermal Physics and Thermal Analysis: From macro to micro highlighting thermodynamics, kinetics and nanomaterials*, Springer Berlin, pp. 237-256.

[10] Šesták, J., (1985). Some thermodynamic aspects of the glassy state. *Thermochim. Acta*, *95*, 459-471, and (1996). Use of phenomenological enthalpy versus temperature diagram (and its derivative-DTA) for a better understanding of transition phenomena in glasses. *Thermochim. Acta*, 280/281, 175-190.

[11] Tammann, G. (1904). About the effect of silicon metatitanic hydrate. *Z. Elektrochemie, 10, 532*, and (1926). The viscosity dependence on

the temperature for supercooled liquids. *Zeit. Anorg. Allgem. Chemie, 156, 245-257.*

[12] Kauzmann, W. (1948). The nature of the glassy state and the behavior of liquids at low temperatures. *Chem. Rev., 43,* 219–256, and Turnbull, D. (1969). Under what conditions can a glass be formed? *Contemporary Physics, 10,* 473-488.

[13] Sakka, S. and Mackenzie, J. D. (1971). Relation between apparent glass transition temperature and liquidus temperature for inorganic glasses. *J. Non-Cryst. Solids, 6,* 145-162, and Zanotto, E. D. (1987). Isothermal and adiabatic nucleation in glass. *J. Non-Cryst. Solids, 89,* 361-370.

[14] Angell, C. A. (1968). Oxide Glasses in Light of the 'Ideal Glass' Concept. I. General Aspects: Ideal and Non- Ideal Transitions. *J. Amer. Ceram. Soc., 51,* 117-124, and (1995). Formation of glasses from liquids and biopolymers. *Science, 267,* 1924-1935.

[15] Hrubý, A. (1972). Evaluation of glass-forming tendency by means of DTA. *Czech. J. Phys. B, 22,* 1187-1193. and (1973). Glass-forming tendency in the GeSx system. *Czech. J. Phys B, 53,* 1263-1272.

[16] Saad, M. and Poulain, M. (1987). Glass forming ability criterion. *Mater. Sci. Forum,* 19 and 20, 11-18, and Poulain, M. (1996). Crystallization in fluoride glasses. *Thermochim. Acta,* 280/281, 343-251.

[17] Weinberg, M. C. (1994). An assessment of glass stability criteria. *Phys. Chem. Glasses, 35,* 119-123, and (1994). Glass-forming ability and glass stability in simple systems. *J. Non-Cryst. Solids, 167,* 81-88.

[18] Lu, Z. P. and Liu, C. T. (2002). A new glass-forming ability criterion for bulk metallic glasses. *Acta Mater., 50,* 3501-3512, and (2003). Glass formation criteria for various glass-forming systems. *Phys. Rev. Lett., 91,* 115505-1/115505-4.

[19] Duan, R. G., Liang, K. M. and Gu, S. R. (1998). A New Criterion for the Stability of Glasses *J. Eur. Ceram. Soc., 18,* 1131-1137.

[20] Ota, R., Wakasugi, T., Kawamura, W., Tuchiya, B. and Fukunaga, J. (1995) Glass formation and crystallization in Li_2O-Na_2O-K_2O-SiO_2, *J. Non-Cryst. Solids, 188,* 136-146.

[21] Cabral, A. A., Fredericci, Jr. C. and Zanotto, E. D. (1997). A test of the Hrubý parameter to estimate glass-forming ability. *J. Non-Cryst. Solids*, *219*, 182-186.

[22] Avramov, I., Zanotto, E. D. and Prado, M. O. (2003). Glass-forming ability versus stability of silicate glasses. II Theoretical demonstration. *J. Non-Cryst. Solids*, *320*, 9-20.

[23] Lu, Z. P. and Liu, C. T. (2002). A new glass-forming ability criterion for bulk metallic glasses. *Acta Mater.*, *50*, 3501-3512, and (2003). Glass formation criteria for various glass-forming systems. *Phys. Rev. Lett.*, *91*, 115505.

[24] Cabral, A. A., Fredericci, Jr. C. and Zanotto, E. D. (1997). A test of the Hrubý parameter to estimate glass-forming ability. *J. Non-Cryst. Solids*, *219*, 182-186, and Nascimento, M. L. F., Souza, L. A., Ferreira, E. B. and Zanotto, E. D. (2005). Can glass stability infer glass forming ability? *J. Non-Cryst. Solids*, *351*, 3296-3308.

[25] Cabral, A. A., Cardoso, A. D. and Zanotto, E. D. (2003). Glass-forming ability versus stability of silicate glasses. I. Experimental test. *J. Non-Cryst. Solids*, *320*, 1-8.

[26] Nascimento, M. L. F., Souza, L. A., Ferreira, E. B. and Zanotto, E. D. (2005). Can glass stability infer glass forming ability? *J. Non-Cryst. Solids*, *351*, 3296-3308.

[27] Li, H. X., Gao, J. E., Wu, Y., Jiao, Z. B., Ma, D., Stoica, A. D., Wang, X. L., Ren, Y., Miller, M. K. and Lu, Z. P. (2013). Enhancing glass-forming ability via frustration of nano-clustering in alloys with a high solvent content. *Sci. Rep.*, *3*, 1983-1/1983-8, and Wu, Z. W., Li, M. Z., Wang, W. H. and Liu, K. X. (2015). Hidden topological order and its correlation with glass-forming ability in metallic glasses. *Nat. Commun*, *6*, 6035-1 6035-7.

[28] Yuan, Z. Z., Bao, S. L., Lu, Y., Zhang, D. P. and Yao, L. (2008). A new criterion for evaluating the glass-forming ability of bulk glass forming alloys *J. Alloys Comp.*, *459*, 251-260, and Long, Z., Xie, G., Wei, H., Su, X., Peng, J., Zhang, P. and Inoue, A. (2009). On the new criterion to assess the glass-forming ability of metallic alloys. *Mater. Sci. Eng. A*, *509*, 23-29.

[29] Kozmidis-Petrovic, A. F. (2010). Theoretical analysis of relative changes of the Hrubý, Weinberg, and Lu–Liu glass stability parameters with application on some oxide and chalcogenide glasses. *Thermochim. Acta*, *499*, 54-60, and (2011). Which glass stability criterion is the best? *Thermochim. Acta*, *523*, 115-23.

[30] Mondal, K. and Murty, B. S. (2005). On the parameters to assess the glass forming ability of liquids *J. Non-Cryst. Solids*, *351*, 1366-1371, and Zhang, P., Wei, H., Wei, X., Long, Z. and Su, X. (2009). Evaluation of glass-forming ability for bulk metallic glasses based on characteristic temperatures. *J. Non-Cryst. Solids*, *355*, 2183-2189.

[31] Du, X. H. and Huang, J. C. (2008). New criterion in predicting glass forming ability of various glass-forming systems. *Chin. Phys. Soc.*, *17*, 249-254, and Guo, S., Lu, Z. P. and Liu, C. T. (2010). Identify the best glass forming ability criterion. *Intermetallics*, *18*, 883-888.

[32] Angel, C. A. (1995). Formation of glasses from liquids and biopolymers. *Science*, *267*, 1924-1935, and (1991). Relaxation in Liquids, Polymers and Plastic Crystals - Strong/Fragile Patterns and Problems. *J. Non-Cryst. Solids*, 131/133, 13–31.

[33] Kozmidis-Petrovic, A. and Šesták, J. (2018). Glass transition temperature its exploitation and new conception of fragility, *Phys Chem Glass*, *Eur J Glass Sci Technol B*, 59, 259-266.

[34] Tanaka, H. (1999). Two-order-parameter description of liquids. I. A general model of glass transition covering its strong to fragile limit. *J. Chem. Phys.*, *111*, 3163-3175.

[35] Angell, C. A, Ngai, K.L, McKenna, G.B, McMillan, P.F and Martin, S.W.(2000). Relaxation in glassforming liquids and amorphous solids. *J. Appl. Phys.*, *88, 3113-3157*.

[36] Angell, C. A. (1985). Spectroscopy Simulation and Scattering, and the Medium Range Order Problem in Glass. *J. Non-Cryst. Solids*, *73*, 1–17.

[37] Angell, C. A. (1988). Structural Instability and Relaxation in Liquid and Glassy Phases Near the Fragile Liquid Limit. *J. Non-Cryst. Solids*, *102*, 205–221 and (1991). Relaxation in Liquids, Polymers and Plastic

Crystals - Strong/Fragile Patterns and Problems. *J. Non-Cryst. Solids*, 131/133, 13–31.

[38] Bohmer, R., Ngai, K. L., Angell, C. A. and Plazek, D. J. (1993). Non-Exponential Relaxations in Strong and Fragile Glass-Formers. *J. Chem. Phys.*, *99*, 4201–4209.

[39] Vogel, D. H. (1921). Rule of the temperature dependence for the viscosity of liquids. *Physikalische Z, 22, 645–646.*

[40] Fulcher, G. S. (1925). Analysis of recent measurements of the viscosity of glasses, *J. Am. Ceram. Soc.*, *8*, 339–355.

[41] Ngai, K. L. (1994). *Disordered effects in relaxation processes.* Springer, Berlin; and Svoboda, R., Málek, J., (2013) Description of macroscopic relaxation dynamics in glasses. *J. Non-Cryst. Solids*, *378*, 186-195.

[42] Kozmidis Petrović, A. (2014). Modified Angell Plot of Viscous Flow with Application to Silicate and Metallic Glass-Forming Liquids. *Int. J. Appl. Glass Sci.*, *5*, 193-205.

[43] Schmelzer, J. W. P. and Gutzow, I. S. (2011). *Glasses and the Glass Transition*, Wiley-VCH, Berlin-Weinheim.

[44] Gutzow, I. S. and Schmelzer, J. W. P. (2013). *The Vitreous State: Thermodynamics, Structure, Rheology and Crystallization.* Springer, Berlin-Heidelberg.

[45] Laughlin, W. T. and Uhlmann, D. R. (1972). Viscous flow in simple organic liquids. *J. Phys. Chem.*, *76*, 2317-2325.

[46] Šesták, J., Kozmidis Petrovic, A. and Živković, Ž. (2011). Crystallization kinetics accountability and the correspondingly developed glass forming criteria. *J. Min Metall B Metall*, *47*, 229–239. Kozmidis Petrovic, A. (2015). Dynamic fragility and reduced glass transition temperature as a pair of parameters for estimating glass forming ability. J. Non-Cryst. Solids, 417/418, 1-9.

[47] Takeuchi, A., Kato, H. and Inoue, A. (2010). Vogel–Fulcher–Tammann Plot for Viscosity Scaled With Temperature Interval Between Actual and Ideal Glass Transitions for Metallic Glasses in Liquid and Supercooled Liquid States. *Intermetallics.*, *18*, 406–411.

[48] Bohmer, R. and Angell, C. A. (1992). Correlations of the Nonexponentiality and State Dependence of Mechanical Relaxations with Bond Connectivity in Ge-As-Se Supercooled Liquids. *Phys. Rev. B.*, *45*, 10091–10094.

[49] Du, X. H., Huang, J. C., Liu, C. T. and Lu, Z. P. (2007). New criterion of glass forming ability for bulk metallic glasses. *J. Appl. Phys.*, *101*, 086106-1/086106-8.

[50] Fan, G. J., Choo, H. and Liaw, P. K. (2007). A new criterion for the glass-forming bility of liquids. *J. Non-Cryst Solids.*, *353*, 102–107.

[51] Du, X. H. and Huang, J. C. (2008). New criterion in predicting glass forming ability of various glass-forming systems. *Chin. Phys. Soc.*, *17*, 249-254.

[52] Yuan, Z. Z., Bao, S. L., Lu, Y., Zhang, D. P. and Yao, L. A. (2008). New criterion for evaluating the glass-forming ability of bulk glass forming alloys. *J Alloys Comp.*, *459*, 251–260, and Fan, G. J., Choo, H. and Liaw, P. K. (2007). A new criterion for the glass-forming ability of liquids. *J. Non-Cryst. Solids*, *353*, 102-107.

[53] Long, Z., Xie, G., Wei, H., Su, X., Peng, J., Zhang, P. and Inoue, A. (2009). On the new criterion to assess the glass-forming ability of metallic alloys. *Mater. Sci. Eng. A.*, *509*, 23–29, and Zhang, P., Wei, H., Wei, X., Long, Z. and Su, X. (2009). Evaluation of glass forming ability for bulk metallic glasses based on characteristic temperatures. *J. Non-Cryst. Solids.*, *355*, 2183–2189.

[54] Cabral, A. A., Cardoso, A. A. D. and Zanotto, E. D. (2003). Glass-forming ability versus stability of silicate glasses. I. Experimental test. *J. Non-Cryst. Solids*, *320*, 1-8.

[55] Guo, S. and Liu, C. T. (2010). New glass forming ability criterion derived from cooling consideration. *Intermetallics*, *18*, 2065-2068.

[56] Illeková, E. and Šesták, J. (2013). Crystallization of Metallic Micro-, Nano-, and Non- Crystalline Alloys, Chapter 13 in book *Thermal analysis of Micro- and Nano- and Non-Crytlline Materials* (J. Šesták, P. Šimon, edts) Berlin, Springer, pp. 257-290.

[57] Hlaváček, B. and Carreau, P. J. (1975). Links between linear and non-linear viscoelatic data for polymer solutions. Chapter 7 in book

Theoretical Rheology (K. Walters, J. F. Hutton, J. R. A. Pearson, editors) Appl. Sci. Publ., London (ISBN 9780853346388); and Hlaváček, B., Šesták, J., (2009) Structural changes in liquids, creation of voids, micromovements of vibrational centers and built-in blocks toward the glass transition temperature. Chapter 18 in book *"Thermodynamic, Structural and Behavior Aspects of Materials Accentuating Non-crystalline States"* (J. Šesták, M. Holeček, J. Málek, edts), OPS-ZČU Plzeň, pp. 388-411, (ISBN 978-80-87269-06-00)).

[58] Hong, L., Novikov, V. N. and Sokolov, A. P. (2011). Is there a connection between fragility of glass forming systems and dynamic heterogeneity/cooperativity? *J. Non-Cryst. Solids*, *357*, 351–356.

[59] Senkov, O. N. and Scott, J. M. (2005). Glass forming ability and thermal stability of ternary Ca–Mg–Zn bulk metallic glasses. *J. Non-Cryst. Solids*, *351*, 3087-3094, and Mondal, K. and Murty, B. S. (2005). On the parameters to assess the glass forming ability of liquids *J. Non-Cryst. Solids*, *351*, 1366-1371.

[60] Nascimento, M. L. F. and Aparicio, C. (2007). Viscosity of Strong and Fragile Glass-Forming Liquids Investigated by Means of Principal Component Analysis. *J. Phys. Chem. Solids*, *68*, 104–110.

[61] Gupta, P. K. and Mauro, J. C. (2008). Two factors governing fragility: stretching exponent and configurational entropy. *Phys. Rev. E*, *78*, 062501-1/062501–3.

[62] Kohlrausch, R. (1854). Theorie des elektrischen rückstandes in der leidener flasche. *Ann. Phys. Chem.*, *167*, 179–214, and Williams, G., Watts, D. C. (1970) Non-symmetrical dielectric relaxation behavior arising from a simple empirical decay function. *Trans. Faraday Soc.*, *66*, 80–85.

[63] Ferry, J. D. (1961). *Viscoelastic Properties of Polymers*. Wiley, New York; and Williams, M., Landel, R., Ferry, J. D. (1955) The Temperature Dependence of Relaxation Mechanisms in Amorphous Polymers and Other Glass-forming Liquids. *J. Amer. Chem. Soc.*, *77*, 3701–3707.

[64] Hlaváček, B., Šesták, J., Koudelka, L. and Mareš, J. J. (2005). Forms of vibrations and structural changes in liquid state, *J. Therm. Anal.*

Calorim, 80, 271-283, and (2002). Mutual interdependence of partitions functions in vicinity of T_g transition *J. Therm. Anal. Calorim., 67*, 239-248.

[65] Ferry, J. D. (1961). *Viscoelastic Properties of Polymers*. Wiley, New York; and Málek, J., Svoboda, R., (2013) Structural relaxation and viscosity behavior in supercooled liquids at the glass transition, Chapter 7 in book *Thermal Analysis of Micro, Nano and Non Crystalline Materials* (J. Šesták, P. Šimon, edits). pp. 147-174, Springer, Berlin (ISBN 978-90-481-3149-5.

[66] Tool, A. Q. (1946). Relation between inelastic deformability and thermal expansion of glasses in its annealing range. *J. Am. Ceram. Soc., 29*, 240–253, and Narayanaswamy, O. S. (1971). A model of structural relaxation in glass. *J. Am. Ceram. Soc*, 54, 491–498.

[67] Moynihan, C. T., Easteal, A. J., DeBolt, M. A. and Tucker, J. (1976). Dependence of the fictive temperature of glass on cooling rate. *J. Am. Ceram. Soc*, *59*, 12-16.

[68] Scherer, G. W. (1990). Theories of relaxation. *J. Non-Cryst. Solids*, *123*, 75–89.

[69] Svoboda, R., Málek, J. and Pustková, P. (2008). Structural relaxation of polyvinyl acetate (PVAc). *Polymer*, *49*, 3176-3185.

[70] Scherer, G. W. (1984). Use of the Adam–Gibbs equation in the analysis of structural relaxation. *J. Am. Ceram. Soc.*, *67*, 504–511.

[71] Šesták, J. (2015). Dynamic cooperative behavior of constituting species at the glass transition vicinity. *J. Therm. Anal. Calorim*, *120*, 167-173.

[72] Chromčíková, M., Liška, M., Lissová, M., Mošner, P. and Koudelka, L. (2013). Structural relaxation of $PbO–WO_3–P_2O_5$ glasses. *J. Therm. Anal. Calorim.*, *114*, 947-954.

[73] Kozmidis-Petrović, A. (2017). The impact of the stretching exponent on fragility of glass-forming liquids. *J. Therm. Anal. Calorim*, *127*, 1975-1981.

[74] Palato, S., Metatla, N. and Soldera, A. (2011). Temperature behavior of the Kohlrausch exponent for a series of vinylic polymers modeled by an all-atomistic approach. *Eur. Phys. J. E*, *34*, 90–95.

[75] Cangialosi, D., Alegria, A. and Colmenero, J. (2009). On the temperature dependence of the non-exponentiality in glass-forming liquids. *J. Chem. Phys*, *130*, 124902-1/-9.

[76] Qin, Q. and McKenna, G. B. (2006). Correlation between dynamic fragility and glass transition temperature for different classes of glass forming liquids. *J. Non-Cryst. Solids*, *352*, 2977–2985.

[77] Martinez, L. M. and Angell, C. A. (2001). A thermodynamic connection to the fragility of glass-forming liquids. *Nature*, *410*, 663–667.

[78] Sastry, S. (2001). The relationship between fragility, configurational entropy and the potential energy landscape of glass-forming liquids. *Nature*, *409*, 164–167.

[79] Dudowicz, J., Freed, K. F. and Douglas, J. F. (2006). Entropy theory of polymer glass formation revisited. I. General formulation. *J. Chem. Phys.*, *124*, 064901-1/–14.

[80] Yue, Y. (2009). The iso-structural viscosity, configurational entropy and fragility of oxide liquids. *J. Non-Cryst. Solids*, *355*, 737–744.

[81] Gupta, P. K. and Mauro, J. C. (2009). The configurational entropy of glass. *J. Non-Cryst. Solids*, *355*, 595–599.

[82] Senkov, O. and Miracle, D. (2009). Description of the fragile behavior of glass-forming liquids with the use of experimentally accessible parameters. *J. Non-Cryst. Solids*, *355*, 2596–603.

[83] Chovanec, J., Chromčíková, M., Liška, M., Shánělová, J. and Málek, J. (2014). Thermodynamic model and viscosity of Ge–S glasses. *J. Therm. Anal. Calorim.*, *116*, 581–588, and Bulíček, M., Málek, J. and Žabenský, J. (2015). On generalized Stokes' and Brinkman's equations with a pressure- and shear-dependent viscosity and drag coefficient, *Nonlinear Anal. Real World Appl.*, *26*, 109-132.

[84] Adam, G. and Gibbs, J. H. (1965). On the temperature dependence of cooperative relaxation properties in glass-forming liquids. *J. Chem. Phys.*, *43*, 139–146.

[85] Avramov, I. and Milchev, A. (1988). Effect of disorder on diffusion and viscosity in condensed systems. *J. Non-Cryst. Sol.*, *104*, 253-260.

[86] Milchev, A. and Avramov, I. (1983). On the Influence of amorphization on atomic diffusion in condensed system. *Phys. Status Solidi B, 120,* 123-130.

[87] Avramov, I. (1991). Influence of disorder on viscosity of undercooled melts. *J. Chem. Phys., 95,* 4439–4443, and (2011). New approach to viscosity of glasses, Chapter 13 in book *Glassy, amorphous and nanocrystalline materials* (Šesták J, Mareš JJ, Hubík P, edts). Berlin: Springer, pp. 59-75.

[88] Mauro, J., Yue, Y., Ellison, A., Gupta, P. and Allan, D. (2009). Viscosity of glass-forming liquids. *Proc. Natl. Acad. Sci. U.S.A., 106,* 19780-19784.

[89] Avramov, I. (2005) Viscosity in disordered media. *J. Non-Cryst. Solids, 351,* 3163- 3173, and (2011). Viscosity, diffusion and entropy: a novel correlation for glasses, Charter 13 in book "*Some Thermodynamic, Structural and Behavioral Aspects of Materials Accentuating Non-crystalline States*" (Šesták J, Holeček M, Málek J, edts) OPS Plzeň (ISBN 978-80-87269-06-0).

[90] Mauro, J. C. and Ellison, A. J. (2011). Breakdown of the Fractional Stokes– Einstein Relation in Silicate Liquids. *J. Non-Cryst. Solids, 357,* 3924–3927.

[91] Zheng, Q., Mauro, J. C., Ellison, A. J., Potuzak, M. and Yue, Y. (2011). Universality of the high-temperature viscosity limit of silicate liquids. *Phys Rev. B, 83,* 212202-1/-5.

[92] Kozmidis-Petrovič, A. (2012). 3D diagrams of equations of viscous flow of silicate glass-forming melts. *J. Non-Cryst. Solids, 358,* 1202–1209.

[93] Sipp, A., Bottinga, Y. and Richet, P. (2001). New high viscosity data for 3D network liquids and new correlations between old parameters. *J. Non-Cryst. Solids, 288,* 166-74.

[94] Smedskjaer, M. M., Mauro, J. C. and Yue, Y. Z. (2009). Ionic Diffusion and the Topological Origin of Fragility in Silicate Glasses. *J. Chem. Phys., 131,* 244514-1/-9.

[95] Angell, C. A. and Hemmati, M. (2013). Glass Transitions and Critical Points in Orientationally Disordered Crystals and Structural

Glassformers: Strong Liquids are More Interesting than We Thought. *Soft Condensed Matter*, arXiv:1302.1645, DOI 10.1063/1.4794546.

[96] Ojovan, M. I. (2008). Configurons: Thermodynamic Parameters and Symmetry Changes at Glass Transition. *Entropy*, *10*, 334-364, and. Schmelzer, J. W. P. and Tropin, T. C. (2018). Glass Transition, Crystallization of Glass-Forming Melts, and Entropy. *Entropy*, *20*, 103; doi:10.3390/e20020103.

[97] Goldstein, M. (2004). Some Thermodynamic Aspects of the Glass Transition: Free Volume, Entropy, and Enthalpy Theories. *J Chem Phys.*, *39*, 3369.

[98] Mareš, J., Šesták, J. and Hubík, P. (2010). Transport Constitutive Relations, Quantum Diffusion and Periodic Reactions. Chapter 14 in book:" *Glassy, Amorphous and Nano-Crystalline Materials"* Springer, Berlin, pp. 227-244 (ISBN 978-90-481-2881-5).

[99] Holba, P., Šesták, J. and Sedmidubský, D. (2013). Heat transfer and phase transition determination at DTA experiments. Chapter 5 in book *Thermal analysis of Micro-, nano- and non-crystalline materials* (J. Šesták, P. Šimon. Edts), pp. 99-134, Springer Berlin (ISBN 9789048131495).

In: An Introduction to Glass Transition   ISBN: 978-1-53615-706-2
Editor: Roberta Ramirez   © 2019 Nova Science Publishers, Inc.

Chapter 2

ATOMISTIC INSIGHT INTO THE GLASS TRANSITION OF LINEAR POLYMERS

Rongliang Wu[*]

Department of Materials Science and Engineering,
Donghua University, Shanghai, China

ABSTRACT

The old problem of glass transition is regarded as the most important behavior of polymer science, while the fundamental microscopic picture of the polymeric glass-formers is scarcely seen. By way of atomistic molecular dynamics simulations, the microscopic local dynamics were analyzed through the conformational transition behavior across a wide range of temperatures. The glass transition temperature could well be predicted through the intersection of separate temperature dependences. Such local dynamics were found, becoming gradually heterogeneous when the temperature went down close to the glassy state. The frozen torsions were determined through a time dependent fashion, and the statistics related to the frozen fraction and frozen chain length showed that the local dynamics of conformational transition corresponded well with the classical thermodynamic or kinetic theories. The frozen chain segments grew as the temperature decreased in a similar way with that of linear condensation

---

[*] Corresponding Author's E-mail: wurl@dhu.edu.cn.

polymerization, and the average frozen chain lengths were related to the configuration entropy during glass transition. The ideal glass transition temperatures, defined in classical thermodynamics theories, were predicted by extrapolating the configuration entropy to zero. The relation between the torsional relaxation time and the configuration entropy showed perfect agreement with the Adam-Gibbs theory around the glass transition temperature. The consecutive frozen torsions had very similar behavior with the cooperative rearranging regions (CRR) defined in the theory. The frozen torsions expanded with a further decrease of temperature until the formation of a volume spanning cluster, which might serve as a premature prototype for the formation of the 'ideal glassy state' with limited accessible configurations.

Keywords: glass transition, conformational transition, CRR, frozen torsion, frozen chain segments, molecular simulation

1. INTRODUCTION

In spite of the wide use of plastics in engineering and everyday life, fundamental knowledge of the details in the glassy condensed state remain poorly understood (Debenedetti and Stillinger 2001; Qiu et al. 2000). The glass transition, which occurs when a metastable liquid is cooled or compressed to form a brittle and amorphous solid, is regarded as the most important behavior of polymeric soft condensed matter (Singh et al. 2013; Angell 1995). Glass transition is marked by a drastic increase of viscosity or relaxation time when approaching the glass transition temperature (T_g), which is commonly regarded as an increase of the activation energy required for chain segment motions (Angell et al. 2000). At high temperatures, where ergodicity prevails, the dynamics are nearly homogeneous, and the dynamic properties usually have a simple Arrhenius behavior with a temperature independent activation energy (Boyd and Smith 2007). Meanwhile, these properties show a marked departure from the Arrhenius behavior in supercooled liquids (Angell et al. 2000; Debenedetti and Stillinger 2001), whose relaxation time is often described by the Vogel-Fulcher-Tammann (VFT) equation (Vogel 1921; Fulcher 1925; Tammann and Hesse 1926). Albeit the continuous efforts in experimental and theoretical studies over the past decades, a predictive molecular theory of glass transition remains to be established (Betancourt et al. 2014), especially the atomistic molecular details associated with the microscopic dynamics.

There have been several phenomenological models for glass transition such as the free volume theory (Fox and Flory 1950). The free volume theory has received substantial data support for its approximate validity for calorimetrically determined glass transition temperature at constant pressure. However, glass transition is also known to occur at constant volume, which is a situation that cannot be well interpreted by the free volume theory. In addition, Maria et al. (Ferrer et al. 1998) showed by analyzing the data in a model-independent manner as a function of temperature and density, and found that temperature is the dominant control variable. The glass transition is not a result of congestion due to a lack of free volume except at very high densities or very high pressures. Ediger's Group (Qiu et al. 2000) came to a similar conclusion that theories based on the free volume effect such as the mode coupling theory (Gotze and Sjogren 1992) are not fundamentally sound, especially at temperatures approaching T_g. The dynamics of the fragile supercooled liquid get strongly heterogeneous (Sillescu 1999), that is, the dynamics in some regions of the sample can be orders of magnitude faster than dynamics in other regions only a few nanometers away (Ediger 2000), and a distribution of local barrier heights can be generated (Qiu et al. 2000).

In the view of thermodynamics, the entropy of a supercooled liquid would eventually come down below that of its stable crystal at a temperature called the Kauzmann temperature, T_K (Kauzmann 1948). But in practice, this entropy crisis is avoided owing to the glass transition phenomenon, at which the structural relaxation time exceeds the timescale of the cooling rate, and further reduction of the entropy intervenes. This suggests that there exists a connection between the kinetics and thermodynamics in glassy materials (Debenedetti and Stillinger 2001). The Gibbs-DiMarzio thermodynamic theory (Gibbs and DiMarzio 1958) states that the configuration entropy goes to zero at temperature T_2, where a unique lowest-energy configuration is formed through an 'ideal glass transition' if the melt is cooled at an infinitely slow rate to assure the attainment of equilibrium. Below T_2, the theory states that the configuration entropy remains zero rather than going to unphysical negative values. Meanwhile, recent computer simulations show that the configuration entropy does not vanish at low temperatures, but reaches a very low plateau value where the particle motions became highly congested with limited accessible configurations (Binder et al. 2003).

The Adam-Gibbs theory (Gibbs and DiMarzio 1958) successfully made a quantitative connection between the large increase of relaxation time in glass-forming liquids and the decrease in the number of configurations that the system is able to sample. The theory suggests that in supercooled liquids, the individual particle motion is frozen and the relaxation can take place only by the cooperative rearrangement of many particles, which is often called the "cooperatively rearranging regions" (CRR). The size or the number of units in these clusters is proportional to the growth of the activation free energy for molecular rearrangement. They further argued that the configuration entropy per CRR, S_c, was almost independent of temperature, and the temperature dependent activation energy involved in molecular relaxation was inversely proportional to S_c. The only weakness of the Adam-Gibbs theory is that it provides no information on the size of CRR as well as its microscopic or molecular description (Debenedetti and Stillinger 2001). Corezzi et al. successfully related the reduction in configuration entropy to the number of chemical bonds formed during epoxy cross-linking (Corezzi et al. 2002). Their findings suggest a nontrivial universality in the relationship of liquid structure to dynamics in dense liquids. This correlation between the thermodynamics and dynamics implies the existence of topological constraints between basin minima and saddle points. They also argued that similar constraints should be able to characterize those regions of configuration space in physical vitrifications and proposed computer simulations for such investigations (Corezzi et al. 2002). In addition, Kanaya et al. (Kanaya et al. 1991) found by quasielastic neutron scattering that the slow process of polymer motion on a time scale from several tens to hundreds of picoseconds could be assigned to an elementary process related to the local chain conformational transitions. Some even pointed out that the glass transition was at least in part due to torsional jumping (Loncharich and Brooks 1990). In the meantime, the atomistic molecular information of the viscous glassy material is difficult to obtain either by scattering or thermodynamic measurements.

With the fast development of computational technology, computer simulations have played important roles in scientific investigations on the microscopic phenomena of macromolecules. Circumfusing the three major features of supercooled liquids (Ediger and Harrowell 2012): the increase in viscosity or relaxation time, the temperature dependence of entropy and the spatially heterogeneous dynamics, varieties of simulation work have been performed on glass transition. Douglas' group (Stukalin et al. 2009) computed the characteristic temperatures of glass transition for poly(α-olefins) based on some coarse-grained models, and estimated the size of the string-like cooperative rearrangements (Shavit et al. 2013). The clusters of highly mobile particles and clusters with relatively low mobility were analyzed and believed to be associated with the diffusive time scale and structural relaxation time scale, respectively (Starr et al. 2013). The string-like cooperative rearrangement clusters observed in coarse-grained polymer melt simulations gave the quantitative relations between the average string length, configuration entropy and the order parameter of the string self-assembly (Betancourt et al. 2014). Ediger et al. (Singh et al. 2013) created ultrastable glasses by means of a computer simulation process that mimics physical vapor deposition. It was found that the extraordinary stability of the glass is associated with some distinct structural motifs and new prospects for understanding glassy materials with superior stability were provided. In addition, the cooperative dynamics of polymers in ultrathin films was also investigated (Peter et al. 2008). Keten et al.(Xia and Keten 2013) used a coarse-grained model of poly (methyl methacrylate) to analyze the influence of substrate adhesion and intermolecular forces on the T_g of thin polymer films. The recent Monte Carlo calculations of Hu et al. (Tang and Hu 2014) investigated the relationship of the sliding motion of short chain fragments with the structural relaxation of ultrathin polymer films, and identified the existence of a new confinement effect at nanoscale.

As mentioned above, the motions of single torsional bonds are of great importance to the microscopic mechanism of polymer glass transition, because their conformational transitions from one rotational isomeric state to another in the polymer backbones are believed to be essential to the dynamics of polymer chains and are closely related to the configuration of the whole system (Kanaya et al. 1999). Takeuchi and Roe (Takeuchi and Roe 1991) as well as Boyd et al.(Boyd et al. 1994) found the conformational transition rates remain Arrhenius in character through T_g with a temperature independent activation energy close to that of a single torsional barrier. Fukuda and Kikuchi (Fukuda and Kikuchi 2000) reported in their simulations of the subglass cis-polyisoprene that more than 50% of the torsional bonds showed no conformational jump during 20 ns MD runs below 300 K, although the relaxation of chain structures or other properties require the ergodic participation of all torsional bonds. The divergence of the time scales between the various macroscopic relaxation processes and the microscopic conformational transition rate was associated with the increasingly heterogeneous character of conformational dynamics with decreasing temperature (Smith et al. 2002). The heterogeneous dynamics was supposed to be caused by the bulk packing, causing bonds to be trapped and oscillate about torsional angles away from the torsional energy minima. In addition, Lyulin and Michels studied the monomer mean square translational displacement of atactic polystyrene in the vicinity of T_g, and found that the local translational mobility was separated into two diffusive regimes by a crossover regime describing the motions confined in cages formed by an almost frozen neighborhood (Lyulin and Michels 2002).

In this chapter, the conformational transition behavior has been utilized to investigate the glass transition behavior of polymers in bulk through atomistic molecular dynamics simulations. The local motion of conformational transition was found to well characterize the glass transition behavior, and the transition barriers in the melt were found to be unexpectedly larger than those in the glass state. The dynamic heterogeneity of local motions was found to be the major cause, and frozen torsions were defined for those with little mobility within different observation times. The statistics associated with these frozen torsions was investigated and a semi-quantitative description of the configuration entropy was derived. The growth of such frozen torsions with a decrease of temperature was compared with the linear growth of polymeric chains during synthesis. Large clusters of such frozen torsions were portrayed and analyzed through their radii of gyration and a premature prototype for the 'ideal glassy state' was proposed. In this way, the atomistic molecular interpretation of polymer glass transition is introduced by way of the microscopic dynamical behavior of the conformational transition.

2. METHODS

2.1. Molecular Dynamics Simulation

Under the assumption of the Born–Oppenheimer approximation that the motion of atomic nuclei and electrons in a molecule can be separated, successive configurations of atomic positions and velocities can be generated through integrating the differential equation of Newton's law of motion (Equ. 1),

$$-\frac{\partial V}{\partial r_i} = F_i = m_i a_i = m_i \frac{dv_i}{dt} = m_i \frac{d^2 r_i}{dt^2} \tag{1}$$

describing the motion of a particle i with mass m_i along the coordinate r_i with force F_i on that particle in that direction. The energy between particles V was written as a function of the nuclear cordinates based upon a simple model of interactions with contributions from bond stretching, angle bending, the rotations about single torsional bonds and the Van der Waals interaction terms (Equ. 2).

$$V = \sum_{bonds} \frac{1}{2} k_b (l - l_0)^2 + \sum_{angles} \frac{1}{2} k_\theta (\theta - \theta_0)^2 + \sum_{torsions} k_\phi [1 +$$
$$\cos(n\phi - \phi_0)] + \sum_{i=1}^{N} \sum_{j=i+1}^{N} \left\{ \frac{q_i q_i}{4\pi\epsilon_0 \epsilon_r r_{ij}} + 4\varepsilon_{ij} \left[\left(\frac{\sigma_{ij}}{r_{ij}}\right)^{12} - \left(\frac{\sigma_{ij}}{r_{ij}}\right)^6 \right] \right\} \tag{2}$$

in which the constants describing the equilibrium bond length l_0 and angles θ_0 as well as their vibrational frequencies k_b, k_θ etc. were parameterized to accurately reporoduce the energetics of high-level quantum chemistry calculations (Smith and Yoon 1994; Byutner and Smith 1999), and the sets of parameters formed the classical force fields for molecular systems.

In this chapter, three different linear polymers were investigated, polyethylene (PE), atactic polypropylene (PP) and the partially fluorinated poly (vinylidene fluoride) (PVDF), and three different chain lengths were investigated for polyethylene, C_nH_{2n+2} (n = 20, 44 and 100), while the other two both had 44 backbone carbons. The force fields used for these polymers have proven accurate in predicting equilibrium and dynamic properties as well as the conformational characteristics of polyethylene (Han et al. 1997) and poly (vinylidene fluoride)(Byutner and Smith 2000). Figure 1 shows the calculated conformational energies of the three polymers with very different styles of energy barriers between the rotational isomeric states, in which the trans states of PVDF split in two and PP has non-symmetric potentials.

The initial random coil configurations were generated from MD of all-trans chains at 600 K in vacuum, and 64 such random coils were then randomly placed into large gaseous cubic boxes with linear dimensions around 20 nm. The initial configurations were energy minimized for thousands of steps before NPT equilibration at 600 K for 20 ns with periodic boundary conditions until the density reached equilibrium. In order to make sure of the relaxation of the chain segments before the annealing process, the characteristic ratios of different chain section length for PE are shown in Figure 2. The characteristic ratios increase steadily with the chain section length until an plateau value around 7, which corresponds well with the experimental value of polyethylene. Meanwhile, shorter chains do not show the plateau regions and are less comparable with the experiments. The equilibrated melt systems were then annealed at the constant rate of 0.1 K/ps to lower temperatures, which span closer around the glass transition region and sparser at higher temperatures. All systems were further equilibrated for 10 ns at their target temperatures under external pressures of 1 bar. The equilibration of these annealed samples was monitored by the density changes, and subsequent production runs were extended for another 10 ns.

All simulations were performed in the GROMACS molecular dynamics simulation package (Hess et al. 2008). The leapfrog algorithm with an integration time step of 1 fs and the trajectories were recorded every 0.1 ps for capturing the fast motion of conformational transitions (Brown and Clarke 1990).

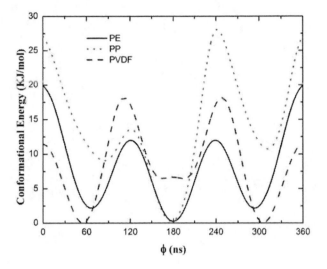

Figure 1. Conformational energies around the backbone torsions of the three polymers investigated.

All bonds were constrained with the LINCS algorithm (Hess et al. 1997) and the Nose-Hoover (Nosé 1984; Hoover 1985) temperature coupling and Parrinello-Rahman (Parrinello and Rahman 1981) pressure coupling methods were used in production runs to control the temperature and pressure, respectively. While in the equilibration runs, the Berendsen method was used for both temperature and pressure for more efficient relaxation. Twin range nonbonded dispersion interactions were truncated between 0.8 and 1.0 nm with the potential shifted for the Van der Waals interactions so that it became exactly zero at the cut-off and made the potential exactly the integral of the force. The long range electrostatic interactions were handled using the Particle Mesh Ewald (PME) method, and the long range dispersion corrections were also implemented for both energy and pressure. The visualization of the molecular structures used the VMD molecular graphics viewer, and the analysis was done using the GROMACS built-in tools or codes written ourselves.

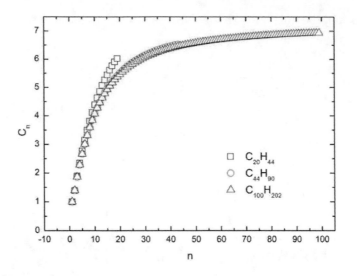

Figure 2. Polymer characteristic ratios averaged over all chain segments along the backbone of the linear polyethylene chains.

2.2. Conformational Transition and the Frozen Chain Segments

Quasielastic neutron scattering results show that the slow process of polymer motion on a time scale from several tens to hundreds of picoseconds could be assigned to an elementary process related to the local chain conformational transitions (Kanaya et al. 1999). The motion of conformational transition around single carbon-carbon bond is hard to capture from experiments, while the atomistic molecular dynamics simulations make it easy to follow such motion of torsional bonds in polymeric systems as shown in Figure 3.

The determination of conformational transition varies in literature, for it depends crucially upon the time interval in recording or comparing subsequent conformations and it also gives different results with different definitions of rotational isomeric states. In addition, the greater the time interval between sampling configurations, the smaller the number of barrier crossings will be counted since the torsions may cross and re-cross back without being detected. It was also demonstrated that the lower limit on the time required for most re-crossings to take place was ~ 0.1 ps (Brown and

Clarke 1990), which is exactly the time interval we chose to keep our trajectories.

As shown in Figure 3, the shallow jumps, defined when the torsion makes a jumps over 20° around the barrier top between different isomeric states, encompass most of the fluctuations, while the deep jumps defined in other works neglect some of the short-lived transitions and many of these short-lived transitions also pass the potential minima. It was also demonstrated that the shallow jumps could well characterize the drastic difference of torsional motions around the glass transition temperature (Liang et al. 2000). Thus, the transition rate $k_{i \to j}$ is defined as the total number of such shallow jumps of conformational transitions N_{ij} from state i to state j divided by the number of dihedrals N_i in state i within a sampling time t.

$$k_{i \to j} = \frac{N_{ij}}{t N_i} = \frac{N_{ij}}{t N \phi_i} \tag{3}$$

in which N is the total number of dihedrals in the system and ϕ_i is the the fraction of dihedrals in state i. The overall transition rate k_t is just the summation of $k_{i \to j}$ between all states within sampling time t. For PVDF, the split two trans states are regarded as a single trans state and the transitions between them are not taken into account, because their values are orders of magnitude larger than others and will inevitably impede the embodiment of other more meaningfull transitions.

The possibility of conformational transition shown in Figure 4 provided more freedom to the system, while those more retarded torsions served like constraints on the whole system. The "frozen torsions" are defined as those with $k_t = 0$ within different sampling time ranging from very short picosecond scale to the whole trajectory length. If one frozen torsion was found, the central position of its two inner atoms in this dihedral would be recorded. If consecutive torsions get frozen, recording of the central positions well prevents any overlap and their automatic connections like the formation of bonds in the visualization programsmake it easier for statistical analysis of frozen chain segments as shown in Figure 12.

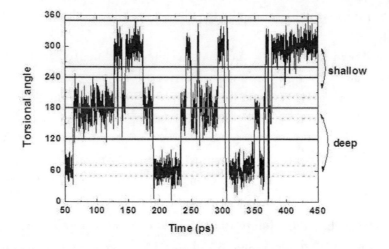

Figure 3. The time evolution of a torsional angle and the definitions for deep and shallow jumps of conformational transitions.

Figure 4. Illustration of frozen torsion in trans state, conformational transition from trans to gauch state with one free room and the coupled transition with two free rooms.

3. RESULTS AND DISCUSSION

3.1. Simulation Structure and Glass Transition Temperatures

The equilibrium specific volumes obtained from the densities of NPT simulations are plotted against temperature in Figure 5. The density of our $C_{100}H_{202}$ is 0.831 g/cm3 at 300 K, which corresponds well with the density of amorphous PE to be 0.855 g/cm^3 at 298 K (Brandrup et al. 1999), for our chains simulated are smaller than the experiments. The solid lines are linear fits, and the intersections of the lines, indicating the volumetric T_g, are located at the temperatures of 204, 243, and 265 K for the $C_{20}H_{42}$, $C_{44}H_{90}$, and $C_{100}H_{202}$ systems, respectively.

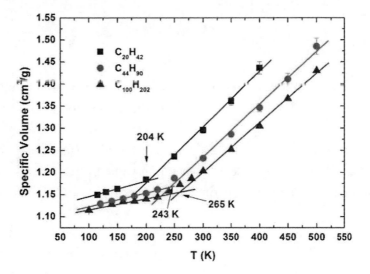

Figure 5. Specific volumes obtained from NPT simulations for the PE systems. The lines are linear fits and the designated intersections are volumetric Tg.

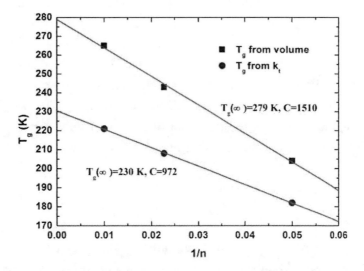

Figure 6. Glass transition temperatures from specific volume and transition rate against reciprocal chain length n. The constant C is typically in the range 10^3-10^4 K mol^{-1}.

The molecular weight dependence of T_g can be described by the equation proposed by Fox and Flory (Fox and Flory 1950) for intermediate chain lengths as shown in Figure 6:

$$T_g(n) = T_g(\infty) - C/n \tag{4}$$

where $T_g(\infty)$ is the transition temperature extrapolated to an infinite chain length of 279 K, which is close to the experimental range of 200 to 250 K depending on the cooling process (Brandrup et al. 2003). The volumetric T_g of PVDF was found to be 340 K which stayed within their experimental range of 233-371 K, while the corresponding value of PP (281 K) was also found to be larger than the experimental range of 253~262 K (Brandrup et al. 2003).

The larger values of the volumetric T_g was considered to be the consequence of the ultra high cooling rate (0.1 K/ps) compared to that used in the experiments. To confirm the equilibrium of conformation in the systems simulated, we examined the conformational distribution of the systems above T_g. Despite the high cooling rate, the trans-conformation fraction increased from ~ 60% in the melt to ~70% around T_g. Making a Boltzmann plot, the logarithmic fraction vs reciprocal temperature gave an energy difference between the trans and the gauche conformation of 2.51 kJ/mol, which was consistent with the experimental value of 2.09 $\pm$ 0.42 kJ/mol and the conformational energy difference of 2.20 kJ/mol in Figure 1. In addition, the single chain form factor of the PE systems were calculated from the simulation trajectories using Equ. 5.

$$P(q) = \frac{1}{N^2} \langle \sum_{i=1}^{N} \sum_{j=1}^{N} \frac{\sin(qR_{ij})}{qR_{ij}} \rangle \tag{5}$$

where N is the number of backbone carbon atoms per chain; R_{ij} is the distance between carbon i and j within the same chain; q is the magnitude of the scattering vector. The angular brackets refer to the average over all chains and time frames of a simulation. The line was calculated as a reference for pure random coils using the Debye function. The $P(q)$ for the systems agrees well with the Debye function at both 500 and 100 K, implying Gaussian behavior of the chains at high and low temperatures, except for the slight deviation for the shorter chains.

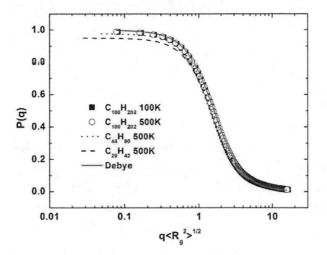

Figure 7. The single chain form factors of polymer chains plotted with the Debye function.

The relaxation of polymer conformation was characterized by the torsional auto correlation functions (TACF) described in Equ. 6.

$$C_\phi(t) = \langle \cos[\phi(t) - \phi(0)] \rangle \tag{6}$$

The angular brackets denote averaging over all time origins and all backbone dihedrals in the system. Such relaxation functions for the longest PE systems are shown in Figure 8. Fitting the auto correlation functions through the Kohlrausch-Williams-Watts (KWW) equation (Kohlrausch 1854; Williams and Watts 1970; Williams et al. 1971)

$$C(t) = \exp\left[-(t/\tau)^\beta\right] \tag{7}$$

and integrating the equation, give the conformational relaxation time of the polymer chains.

$$\tau_\infty = \int_0^\infty \exp\left[-(t/\tau)^\beta\right] = \frac{\tau}{\beta}\,\Gamma\left(\frac{1}{\beta}\right) \tag{8}$$

The relaxation time of the systems at different temperatures fit well with the Vogel-Fulcher-Tammann (VFT) equation (Vogel 1921; Fulcher 1925; Tammann and Hesse 1926).

$$\tau_\infty = A\exp\left(\frac{B}{T-T_0}\right) \tag{9}$$

which is equivalent to the well-known Williams-Landel-Ferry (WLF) equation for polymer materials (Williams et al. 1955) on the assumption that the vibrational entropy of the supercooled liquid and its stable crystal is the same (Debenedetti and Stillinger 2001). T_0 is believed to be close to T_K, which indicates a connection between kinetics and thermodynamics not present at higher temperatures (Williams et al. 1955; Schönhals et al. 1993). Fitting the torsional relaxation data, except the last two values, gives $T_0 = 138$ K, which is slightly larger than the value of 122 K in the work of Douglas et al. (Stukalin et al. 2009)

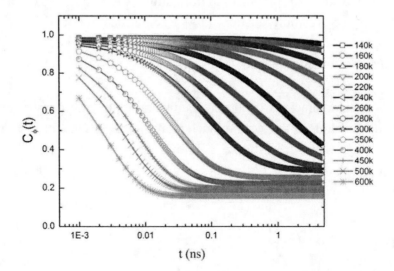

Figure 8. The torsional auto correlation functions of the longest PE systems.

3.2. Conformational Transition Behavior across T_g

Conformational transition is of local dynamics in polymer systems. The transitions between the rotational isomeric states occur to each backbone torsion and the accumulation of such conformational transition is believed to be the very origin of the global changes in polymer chain systems. The overall transition rates, k_t, are plotted in logarithm with respect to inverse temperatures for all systems in Figure 9, which shows perfect Arrhenius relationships with temperature in the high temperature region.

$$k_t = Ae^{-(E_a/RT)} \tag{10}$$

where E_a represents the activation energy or barrier for conformational transition and R is the ideal gas constant.

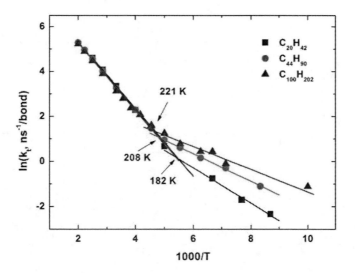

Figure 9. The overall transition rates against inverse temperature of amorphous PE.

When the temperature decreases through T_g in the glass state, significant deviation of k_t from the Arrhenius line is observed. In the low temperature region, the rates also fit well with the Arrhenius equation for all chain lengths. Such inflections of transition rates could not be fit with the Vogel–

Fulcher equation introduced above. The intersections of the transition rates take place below the volumetric T_g, located at 182, 208, and 221 K for the $C_{20}H_{42}$, $C_{44}H_{90}$, and $C_{100}H_{202}$ systems, respectively. The corresponding values for PP and PVDF are 260 and 310 K, respectively. These T_g values can also be well fit by eq 4, which yields a $T_g(\infty)$ value of 230 K (see Figure 6), which is closer than the volumetric $T_g(\infty)$ to the experimental T_g values within (200-250 K). It seems that the T_g obtained from the conformational transition rates is more accurate than that from the simulated volumes, thus the conformational transitions may exhibit the fundamentals molecular nature of glass transition.

Table 1. Activation Energies of Conformational Transitions

		E_a (kJ/mol)	
		$<T_g$	$>T_g$
$C_{20}H_{42}$	overall	6.48	12.49
	g→t	6.99	11.09
	t→g	6.39	13.79
	g↔g	8.61	17.59
$C_{44}H_{90}$	overall	4.96	12.51
	g→t	5.02	11.11
	t→g	4.93	13.30
	g↔g	10.55	17.73
$C_{100}H_{202}$	overall	3.95	12.53
	g→t	3.97	10.98
	t→g	3.94	13.43
	g↔g	12.46	17.93
PP	overall	3.84	11.15
PVDF	g→t	9.32	16.34
	t→g	9.10	15.87
	g↔g	10.95	15.39

The activation energies or the transition barriers obtained from Equ. 10 for the overall and specific transitions between the RIS are listed in Table 1. Results show that above T_g, the transition barriers are very close to each other, which indicates that the conformational transition behavior at high temperatures are nearly the same for all chain lengths. The barrier differences between g→t and t→g values are very close to the potential energy difference, about 2.20 kJ/mol, seen in Figure 1, and the results are reversed for PVDF with the gauche state having the lowest energy. The barriers for g↔g transitions are the highest of the three, with a value near 18 kJ/mol, which is 2 kJ/mol lower than the one from the potential energy. These results indicate that the barriers obtained from the kinetic process on the free energy surface are different from those from the potential curve, but close to each other. Below T_g, all barriers were found drastically reduced. This finding is unexpected but truly exists. The barrier height in the glass state is lower than that in the melt state. This phenomenon was found here related to the chain being "frozen," the breaking down of ergodicity in the glass state. The break down of ergodicity leads directly to the inhomogeneous behavior.

3.3. Heterogeneous Distribution of Conformational Transition

The spatial heterogeneity in dynamics has been evidenced to exist in polymer and super-cooled liquids near T_g (Ediger 2000). The heterogeneity means the local dynamics in some regions of the sample can be orders of magnitude faster than the dynamics in other regions only a few nanometers away. The conformational transition behavior also exhibits such heterogeneous distributions, since it is of local dynamics. Figure 10 shows the overall transition rates of each bond in the $C_{44}H_{90}$ system along the abscissa axis. In Figure 10a, the conformational transition rates are relatively homogeneous and they fluctuate around an average value above T_g. Figure 10b describes the transition behavior in the glassy state, and the transition rate are highly heterogeneous: some increased tremendously, even higher than those above T_g, and some greatly decreased. It is evident that many bonds have $k_t = 0$ in the curves at below T_g, as was also found by Fukuda and Kikuchi (Fukuda and Kikuchi 2000). The number of such frozen bonds increases as the temperature decreases, making certain restraints on the whole system as illustrated in Figure 4. On the contrary, those activated torsions should make major contributions to the phenomena of a larger k_t and reduced barrier in the glass state. The details on these transitions might offer more information on how these activated torsions are generated.

The details of the conformational transition behavior can be characterized by the departure dihedral and the transition dihedral shown in Figure 11. The departure dihedral is the dihedral site, where the dihedral starts a transition. The transition dihedral is the region the dihedral sweeps over during a transition. Figure 11a shows the distributions of the departure dihedrals during the g→t transitions at all temperatures for the $C_{100}H_{202}$ system. At 300-500 K, the departure dihedral distributions have almost the same curves, spanning nearly the entire basin of RIS. While below T_g, a significant decrease of population around the basin center (~ 69°) and a sharp enhancement of population near the barrier top (120°) can be distinctively seen.

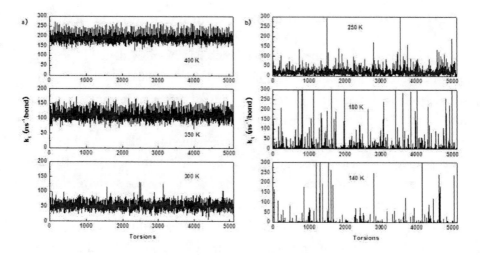

Figure 10. The transition rate of each torsional bond of the $C_{44}H_{90}$ system (a) in the melt and (b) in the glass state.

The transition dihedral distribution given in Figure 11b shows a similar phenomenon. Though a broad definition of RIS (100°) is used in our shallow jumps, the fast oscillations around the barrier top (with transition dihedrals ~20°) do not prevail at all temperatures. At high temperatures, the dihedrals cross over a broad range of dihedral angles from 20° to more than 100°; while below T_g, most dihedrals sweep over $30 \pm 10°$ and many of the larger magnitude transitions from $60° \sim 100°$ are inhibited.

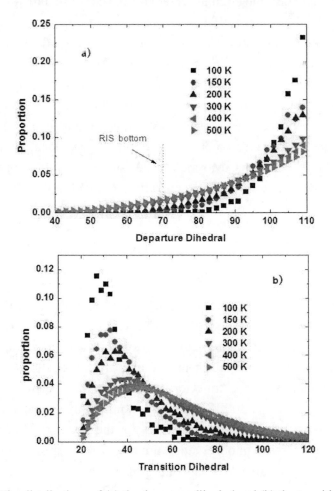

Figure 11. The distributions of (a) the departure dihedral and (b) the transition dihedral of g→t transitions at all temperatures for the $C_{100}H_{202}$ system.

As already found in Figure 10b, below T_g the local dynamics is substantially activated and some torsions have transition rates, orders of magnitude higher than expected. These activated torsions are contributed from small magnitude transitions, and an activated dynamics mechanism was already assumed (Boyd and Smith 2007). They thought that in glasses where the dynamics get severely retarded, overpopulation of torsions divergent from the potential minima are located, and the residence times at these sites are so long that the bonds can be considered to be trapped. These torsion sites may become eligible centers for the conformational transitions, which result in the reduced effective barrier.

Our result of the departure dihedral and the transition dihedral indicates that in the glass state, the conformational transition behavior becomes (1) starting very close to the top of the barrier, (2) jumping a relatively small step, about 30°, which supports the above mechanism. It is apparent here that the number of activated torsions is still quite minor seeing from Figure 10, although they make major contributions to the local dynamics of conformational transitions. Most of the torsions are still trapped within different isomeric states, which is generally attributed to the "frozen" local conformational structure or environment.

3.4. Frozen Torsions Indicating Glass Transition

The frozen torsions capture the time dependence of the local dynamics and give a more detailed and direct overview on the glass transition as demonstrated in Figure 12 for the $C_{100}H_{202}$ system. The left panel shows the frozen torsions at the indicated temperatures of 160 K, 200 K and 240 K observed within a 6 ns trajectory, and the right panel shows the corresponding results at 300 K observed within relatively short time scales of 20 ps, 100 ps and 500 ps, respectively. For the frozen torsions shown in the figure, the increase in observation time has a very similar effect to the increase of temperature. At temperatures far above T_g or within a long observation time, no torsions get frozen. With the decrease of temperature or observation time, the frozen torsions emerge. In the beginning, the frozen torsions are sparsely populated in the system and very few consecutive torsions are present. Further decrease of temperature or observation time leads to more frozen torsions, and the consecutive ones connect in a chain-like fashion. Finally, when the temperature is low enough or within a very short observation time, these frozen torsions percolate the entire system.

More quantitative information can be obtained through the frozen fraction α, which is defined as the ratio between the number of frozen torsions and the total number of torsions in the system $\alpha = N_{frozen}/N_{dih}$. All frozen torsions collected at the target temperatures and different observation times are shown in Figure 2a. Within a very short observation time, such as 20 ps or 50 ps, nearly all torsions get frozen at the low temperature end; while at the high temperature end, nearly all torsions are able to perform transitions between different RIS if sufficient observation time is provided. The variation of the frozen torsions with the increase of temperature exhibits similar behavior with relaxations, and all data could be well fit by the empirical equation,

$$\alpha\,(T) = \exp\left[-\left(T/T_g\right)^{\beta}\right] \tag{11}$$

in which T_g and β are fitting parameters and T_g represents the temperature when the frozen fraction is reduced to $1/e$. The convergence of T_g values was observed, for they move closer to each other at longer observation times. The reason why we use this fitting parameter as T_g is that it coincides well with the microscopic T_g obtained from conformational transition rates as illustrated in Figure 9, which provides a microscopic T_g of 221 K for a 2 ns trajectory and the 10 ns trajectory provides a microscopic T_g of 206 K, and both values are very close to the temperatures at which the frozen fractions decrease until 1/e. In addition, the converging value of T_g around 195 K corresponds well with the experimental value and other simulation work of 196 K (Stukalin et al. 2009). These results show that when a certain fraction, namely $1/e$, of torsions get frozen, the conformational transition behavior of torsions in the system changes due to the constraints imposed by these frozen torsions, which results in the dynamical heterogeneity of these local dynamics or the activation and deactivation of some torsions. The activated torsions are recorded by the overall transition rates in Figure 9 with smaller transition barriers, and the deactivated torsions are recorded by the frozen torsions in Figure 13 over different observation times. As for the volumetric T_g obtained from the transition of density around 250 K in Figure 5, we find it related to the emergence of the frozen torsions between 240~260 K. Thus, when torsions start to get frozen, the density of polymers start to deviate from the high temperature line.

The kinetics of the conformational transitions can well explain the above phenomena. Because all torsions stay in their local potential valleys on the potential energy surface, at high temperatures they had enough energy to overpass those barriers surrounding them, making their behavior ergodic; At low temperatures, some torsion may be locked in valleys with relatively low barriers, while others may have local barrier heights which are too high for them to overpass at this temperature. If a very short observation time is provided, even those torsions with relatively low transition barriers are regarded as frozen torsions because they do not have enough time to fulfil the transition.

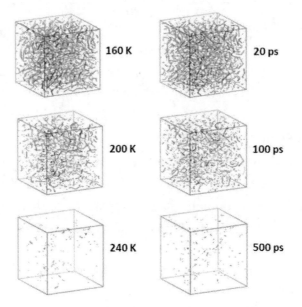

Figure 12. Frozen torsions in the backbone of the polymers at selected temperatures below Tg within a 6 ns observation time (left); and frozen torsions at 300 K within different observation times (right).

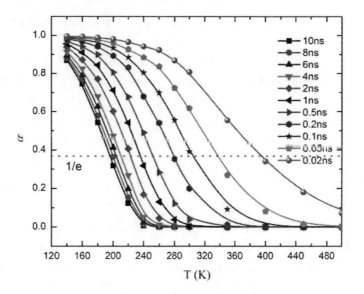

Figure 13. Variation of the frozen fraction α with temperature within different observation times.

If we increase the observation time, more and more torsions have the opportunity to jump out of their valley, but there will always be some torsions deeply trapped at their corresponding temperatures. These trapped torsions in deep valleys are collected by the frozen fractions, and this value will eventually converge at longer observation times as illustrated in Figure 13. As far as the conformational transitions are concerned, the observation time can be regarded as something in quality similar to the annealing rate in experiments. A very fast annealing rate would give only very short action time for local dynamics, and a slower annealing rate would give longer action time at the target temperatures. Thus, the frozen torsions obtained within a relatively short observation time could also be a good measure of the microscopic nature of the glassy state. On the contrary, within a relatively long period of 10 ns for the fast conformational transition process, there are still 90% of frozen torsions at the temperature of 140 K. In this way, the heterogeneous nature of the local dynamics can be well represented by the time dependence of the frozen torsions.

3.5. The Frozen Chain Length and the Configuration Number

As can be seen from Figure 12, the frozen chain segments show similar behavior with that of linear polymerization. It was found in chemical vitrification (Corezzi et al. 2002) that the formed covalent bonds between monomers were expected to impose configuration restrictions and force the molecules to move cooperatively over an increasing length scale as the reaction proceeded. It was also proposed that similar restrictions should also exist in physical vitrifications waiting for computer simulations to unveil (Corezzi et al. 2002). The frozen fractions α, defined above, can be regarded as something similar with the reaction extent or the fraction of reacted functional groups during the polymer synthesis. They also make restrictions on the conformational transitions since the mobility of torsions around these frozen ones gets greatly reduced, and thus the configuration numbers should change accordingly. For our linear polymer chains, the growth of consecutive frozen torsions is completely random and never forms loops or networks, which makes it reasonable to see the freezing process as linear polymerization. The variation of the average number of frozen chain lengths with observation time is illustrated in Figure 14, in which we have also provided data collected within very short observation times. The long frozen chain segments are only formed at low temperatures or within short observation times and the increase of the frozen chain length has clear power law dependence with the observation time. The dependence seems to change around the observation time of 0.1 ns, below which the frozen torsions captured might not be representative of the local dynamics because the time period is too short. In the following sections, results obtained over more than 0.1 ns will be the focus unless specifically noticed.

According to the approximations in the work of Corezzi (Corezzi et al. 2002), who made a similar connection between the configuration entropy and the chemical bonds formed during chemical vitrification of network epoxy polymers, the number of configurations at the conversion rate or frozen fraction α have:

$$\Omega_c(\alpha) = \Omega_c(0)^{1/x_n(\alpha)} \tag{12}$$

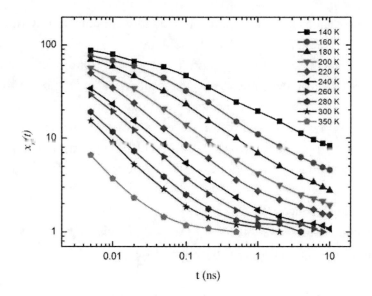

Figure 14. Variation of the average number of frozen chain lengths with observation time at various tempearatures.

where $x_n(\alpha)$ is the average degree of polymerization at conversion α, or the average frozen chain length at the frozen fraction α in the present work. In chemical vitrification, two monomers linked with each other are expected to behave approximately as a single cooperative unit (Corezzi et al. 2002); while in our simulated polymer system, a single frozen torsion already represents the motion of 4 backbone atoms. According to the linear polymerization without branching we have $x_n(\alpha) = 1/(1 - \alpha)^\delta$ ($\delta = 1$ for linear condensation), in which $x_n(\alpha) = 1$ when $\alpha \to 0$. At the very beginning of the process ($\alpha \to 0$), the initial value for $x_n(0) = 1$ can also be seen from Figure 14. The initial value of chain length means that the freezing process starts when a very small number of individual torsions starts to get frozen, just like what is illustrated in the lower panel of Figure 9. For systems at high temperatures and within long observation times $x_n(\alpha) = 0$, the 'chain growth' process has not started yet and this restriction function does not apply. Thus, during the freezing process $x_n(\alpha)$ does not go to zero.

The average number of frozen chain lengths are plotted in double logarithm with $1/(1 - \alpha)$ in Figure 15 using all data except those within very short observation times. The grids show different power law dependence regions and the δ values indicated were obtained by fitting the power law equation. Below the length of 20, the value of δ = 0.953 ± 0.010 shows good linear dependence. In addition, the inset figure with linear axes also shows the linear dependence clearly within the same region. Above the length of 20, the power law dependence gradually moves to another region with δ = 0.496 ± 0.018, and then to an even smaller δ, which is attributed to the relatively short chain length of 100 in our simulations. In real polymer glasses with much longer chains the linear relationship should be able to last until a longer chain length. Even with the relatively short chains simulated, the linear dependence still overwhelms, and for that, when $1/(1 - \alpha) = 20$ we need $\alpha = 0.95$ and this frozen fraction already includes most part of the torsion freezing process as can be seen from Figure 13.

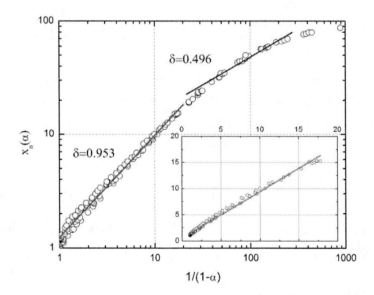

Figure 15. Scalar dependences of the average number of frozen chain lengths on the frozen fraction. The δ values denote the power law fitting results and the inset show the initial linear dependence, which persists until 17.

The linear dependence of the average number of frozen chain lengths coincides with the processes of one dimensional percolation or the linear condensation polymerization, and the 0.5 power law dependence may be caused by the association of these frozen chains, which is similar with the results of the freely association model proposed in the work of Douglas et al. (Betancourt et al. 2014).

3.6. Configuration Entropy and Glass Transition Temperature

Under the athermal approximation, the configuration entropy can be estimated from the average number of frozen chain lengths through the definition of entropy,

$$S_c(\alpha) = k_B \ln\Omega_c(\alpha) = S_c(0)/x_n(\alpha) \tag{13}$$

where $S_c(0) = k_B \ln\Omega_c(0)$ is the configuration entropy at the very beginning of the process with $\alpha \to 0$, where almost all torsions could perform a conformational transition between RIS. Using the relationship between $x_n(\alpha)$ and α in Figure 15, we may relate the the relative configuration entropy to the frozen fraction through $s_c = S_c(\alpha)/S_c(0) = (1 - \alpha)^\delta$, which measures the decrease of entropy with the increase of frozen torsions. A very similar relation between the configuration entropy and the 'extent of polymerization' was also found in the string model of glass-forming liquids, in which their free association model gave $\delta = 0.5$ while the living polymerization model gave $\delta = 1$. (Betancourt et al. 2014) The relative entropy with respect to the temperature reduced by T_g is illustrated in Figure 16.

All frozen fractions within different observation times seem to abide by a universal relation, and at high temperatures the configuration entropy comes to a plateau value of $S_c(0)$. The red line in Figure 16 is the fitting of all data by the expression for the relative configuration entropy,

$$s_c = 1 - \alpha = 1 - \exp\left[-(T/T_g)^\beta\right] \tag{14}$$

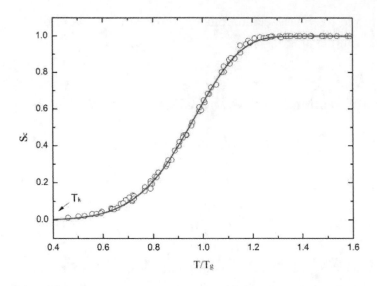

Figure 16. Relative configuration entropy with respect to the reduced temperature, and the red solid line is the fitting of Equ. 14.

in which $\beta = 6.48$ and T_g is the glass transition temperature obtained from Figure 13. Around $T = 0.4\ T_g$, the configuration entropy almost disappears, which should correspond to the 'ideal glass transition' temperature T_2 or T_K. (Gibbs and DiMarzio 1958; DiMarzio and Gibbs 1958) To compare with the experimental results, the volumetric $T_g = 279\ K$ was predicted for amorphous PE with infinite chain length in previous sections, and the ideal glass transition temperature should occur ~ 112 K, which corresponds well with the value of 110 K obtained by Douglas et al. (Stukalin et al. 2009).

3.7. The Dynamics of the Frozen Chain Segments and the CRR

The phenomena of gradual evolution of properties to their equilibrium values over time at temperatures below T_g are called "physical aging." Theoretically, if an infinite observation time was provided, all configurations could be well sampled. In practice, the properties of a polymer glassy material depend on the process by which it is formed. Specifically, T_g increases with the cooling rate because the slower the liquid is cooled, the longer the time is available to sample more configurations (Debenedetti and Stillinger 2001). We may invert the data in Figure 13 to obtain the time dependence of frozen fractions in Figure 17, which gives an exact picture of physical aging or the dynamics of chain segments at different temperatures. In order to have more data for those at high temperatures, we have added more data at shorter time scales. Above T_g, nearly all chain segments were able to get completely relaxed within the 10 ns trajectory, the relaxation time can be as short as 0.1 ns for those at relatively high temperatures. A longer time is needed for complete relaxation of the chain segments approaching T_g, and below T_g there always exist torsions that do not perform any transition between their RIS, which indicates the freezing of local chain segments. The complete relaxation of even such small and local motions need much longer time than our trajectory length of 10 ns.

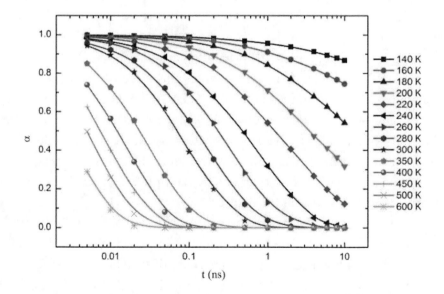

Figure 17. Transposed data of the frozen fractions give the time dependence of the frozen fractions.

In the Adam-Gibbs theory (Gibbs and DiMarzio 1958), the structural relaxation time τ and the configuration entropy S_c are related as:

$$\tau = \tau_0 \exp(C/Ts_c) \tag{15}$$

in which τ_0 and C are temperature independent constants. The continuous slowing down of dynamics or the increase of relaxation time is related to the loss of configurations in the system, which is expressed by the cooperative motions in the system. The lower the configuration entropy s_c is, the greater the size of the cooperative rearranging regions will be. This relation corresponds well with our results of frozen chain length described in Equ. 13. The relationship between the torsional relaxation time obtained from Figure 8, $-\ln\tau$, and $1/Ts_c$ is plotted in Figure 18, in which the entropy values are obtained within different observation times and the same relaxation time is used for the respective systems since there exists only one relaxation time at a certain temperature. In this way, we may check whether the values obtained within short observation times could well represent the glassy state or abide by the Adam-Gibbs theory. At the high temperature end, where the configuration entropy gives the value of unity, the plot is identical with that illustrated in the inset figure. With the increase of frozen chain segments within shorter observation time or at lower temperatures, $1/Ts_c$ show perfect linear relationship with logarithmic relaxation time for curves. The temperature at which this relationship holds starts earlier at higher temperatures within shorter observation times as was also shown in Figure 13, where the frozen fractions emerge.

The CRR defined in the Adam-Gibbs theory (Adam and Gibbs 1965) are cooperative units with specific configuration entropy, and the activation energy for relaxation is extensive in the mass of the CRR. It was found in the work of Douglas et al. (Betancourt et al. 2014) that the string length determined from their coarse-grained simulations was the most quantitatively valid choice for CRR. In this chapter we have suggested the atomistic frozen chain length from the consecutive frozen torsions. The average number of frozen chain lengths has $x_n(\alpha) = 1/1 - \alpha$ before α reaches 0.95, where $s_c = 1/x_n(\alpha)$. In this way, we may infer from Equ. 15

or Figure 18 that during the glass transition $-\ln\tau \sim x_n(\alpha)$, which demonstrates the frozen chain segments determined from our frozen torsions also abide by the CRR assumption. In this way, we have proposed a probable atomistic point of view over the CRR regarding the glass transition. When the torsions get frozen in the chains, these frozen chain segments have to move cooperatively by dragging or sliding of other units, and within these frozen chain segments motions are limited to bond or bend vibrations around their equilibrium values.

In fact, there is deviation from the linear relationship at even lower temperatures, which could be estimated from the inset figure that the last two data show a distinct deviation at the low temperatures. Possible reasons for this deviation might be the deviation from the linear relationship at high frozen fractions and the inaccuracies of the relaxation times calculated at low temperatures. Another possible reason will also be analyzed in the next section.

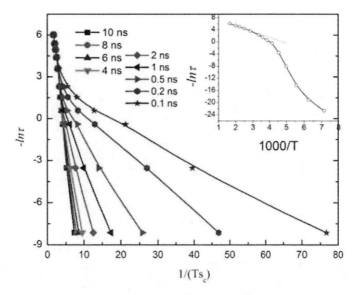

Figure 18. Verification of the Adam-Gibbs relationship for the relaxation time of the frozen chain segments and the configuration entropy (the three data at the low temperature end were not shown). The timescales indicated, show the trajectory length used for the determination of configuration entropy and the inset figure shows the Arrhenius relationship of relaxation time at high temperatures.

3.8. The Max Frozen Clusters and the Volume Spanning Clusters

More detailed structures of these frozen chain segments can be obtained from their radial distribution function in Figure 19, in which apparent short range correlations exist for these frozen torsions. The first peak corresponds to the adjacent torsions, and the two split peaks between 0.2 and 0.3 nm correspond to torsions in different conformational states spaced by single non-frozen torsion. The two peaks between 0.3-0.4 nm correspond to torsions spaced by more activated torsions. The distinction between short range and longer range correlations were chosen at 0.4 nm, where the strong short range correlations seem to diminish. In subsequent analysis, the frozen chain torsions spaced less than 0.4 nm are regarded as a single cluster instead of the consecutive torsions described in previous sections.

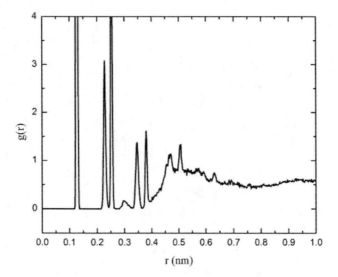

Figure 19. Radial distribution function of the frozen torsions shows the short range correlation between these segments rests ~ 0.4 nm.

The largest clusters at the selected temperatures are shown in Figure 20, which exhibit the 3-dimensional correlations between these frozen torsions and similar effects of observation time and temperature are also observed.

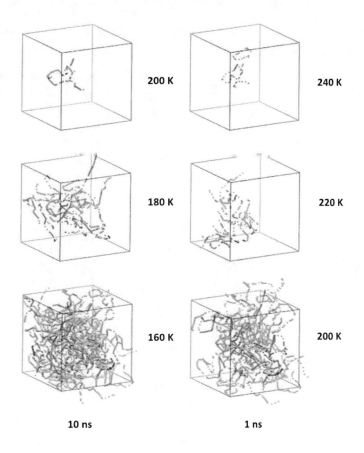

Figure 20. The max frozen clusters observed within 10 ns and 1 ns at the indicated temperatures.

At relatively higher temperatures, only regional clusters of frozen torsions in single or adjacent chains are formed. As the temperature further decreases, more such local frozen chain segments get correlated and the max cluster expands. Until a certain temperature such as the 160 K for the 10 ns trajectory and the 200 K for the 1 ns trajectory, the max cluster spans the entire system and further decrease of temperature should not be able to change the overall behavior of the whole system because the system seems to be frozen out by this max cluster.

To get a general view of the effects of clusters on glass transition, we have plotted the radius of gyration (R_g) of the max clusters as well as the average values of all such clusters with the reduced temperature in Figure

21a. Both data fit well onto similar trends albeit the clusters are formed within different observation times. The average R_g shows an apparent transition at around 0.8 T_g, above which both sizes decrease linearly with temperature.

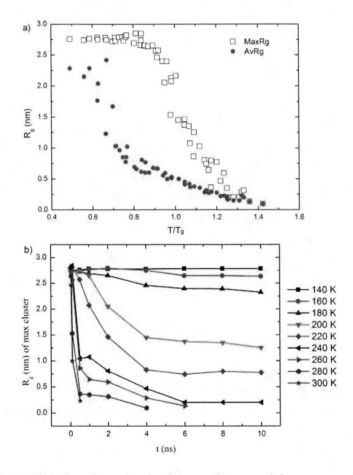

Figure 21. a) The radius of gyration for the max clusters and the average value for all clusters in the system plotted with the reduced temperature; a) the evolution of the radius of gyration for the max clusters in the systems at different temperatures with the observation time.

Below 0.8 T_g, the size of the max clusters levels off at ~2.75 nm, which is exactly half the size of our system, showing a volume spanning cluster percolate the entire system. In the meantime, the average Rg still increases with the decrease of temperature owing to the growing of other chain segments within the framework of the percolating max clusters. In addition, the transition temperature of 0.8 T_g also coincides with the deviation from the Adam-Gibbs relationship in Figure 18. Consequently, the volume spanning max clusters formed at the low temperature end might have restricted further relaxation of the whole system structure.

The evolution of the T_g for these max clusters is plotted in Figure 21b with the observation time. Above Tg, the size of the max cluster decreases very fast with the observation time, which corresponds well with the frozen chain length in Figure 3. Within short observation times at the lower temperature end, all max cluster show similar size of around half the system box size, namely they are percolating the entire system because of the periodic boundary conditions used in our simulations. The volume spanning cluster in the system serves as a framework preventing the further decrease of the configuration entropy, so it might well serve as a premature prototype for the formation of the 'ideal glassy state' with limited accessible configurations (Binder et al. 2003). Thus, the entropy of the system starts to level off at the low temperature end as indicated by the Gibbs-DiMarzio theory (Gibbs and DiMarzio 1958; DiMarzio and Gibbs 1958). In addition, the formation of such percolating clusters might be one of the possible reasons causing the deviation from the Adam-Gibbs relationship in Figure 18 at the lower temperature end, since in the theory it was argued that the supercooled fluid can be decomposed into a lot of CRR, each having a certain configuration entropy.

In order to show that the volume spanning cluster is not due to the limitation of the system size, we have provided the R_g evolution of a larger system ~7 nm with 125 chains of the same type in Figure 22 in comparison with the smaller system at 200 K, the two curves show similar behavior with the observation time and within smaller observation times both systems formed volume spanning frozen clusters. At intermediate temperatures around T_g, the max cluster sizes shrink with the increase of observation time to an equilibrium plateau value, which is quite regional since they are much smaller than half the system size. Within this temperature range, the frozen

Rongliang Wu

clusters might also serve as a good candidate for the CRR in a larger scale than the consecutive frozen torsions described in previous sections.

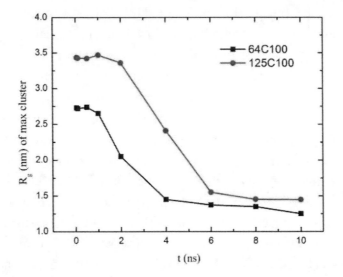

Figure 22. Variation of the max cluster size at the same temperature of 200 K for 125 $C_{100}H_{202}$ and 64 $C_{100}H_{202}$, whose system sizes are around 7.0 nm and 5.5 nm respectively. The frozen torsions obtained within smaller observation times in both systems formed volume spanning frozen clusters.

CONCLUSION

By way of the most microscopic dynamic site in linear polymers, namely the conformational transition around the backbone bonds, we have successfully established a connection between glass transition and the frozen torsions. The fraction of these frozen torsions gives the picture of the most microscopic local dynamics in polymer systems, and the frozen fractions seem to converge at longer observation times. The emergence of the frozen torsions is found to be a possible cause of the volumetric glass transition, where the structure of the supercooled liquid starts to get jammed. The T_g obtained from the conformational transition rates corresponds to the temperature at which the frozen fractions are reduced to 1/e from almost unity. The growth of the frozen torsions with the decrease of temperature can be seen as a linear polymerization process or the one dimensional percolation, and the 'degrees of polymerization' $x_n(\alpha) = 1/(1 - \alpha)$ before these frozen chain segments grows longer than 20 torsions. Later deviation from the linear relationship can be regarded as the mutual meeting of these frozen chain segments as well as the restriction of the relatively short polymer chains simulated in the experiment. In reality, this relationship is believed to exist until a much longer length scale for engineering polymers Based on the estimation of the number of restrictions from the frozen torsions, the relative configuration entropy is deduced to be the inverse of the number average frozen chain length. The ideal glass transition temperature where the configuration entropy is extrapolated to zero is found to lie within the reasonable range with those reported in literature. Physical aging or relaxation of the frozen chain segments gives the relaxation time, which shows perfect linear relationship with inverse temperature at high temperatures described by the Arrhenius relation, but the Adam-Gibbs equation is abided by at the lower temperature range. At even lower temperatures, the frozen torsions form volume spanning clusters with R_g of the max cluster to be half the system size. The temperature at which the max cluster percolates the entire system coincides well with the deviation of the Adam-Gibbs relation, and the volume spanning cluster formed through these frozen torsions might serve as a potential framework for the ideal glassy state at lower temperatures.

REFERENCES

Adam, Gerold, and Gibbs, Julian H. 1965. On the Temperature Dependence of Cooperative Relaxation Properties in Glass-Forming Liquids. *J. Chem. Phys.* 43 (1):139 -146.

Angell, C. A. 1995. The Old Problems of Glass and the Glass Transition, and the Many New Twists. *Proc. Natl. Acad. Sci.* 92:6675-6682.

Angell, C. A., Ngai, K. L., McKenna, G. B., McMillan, P. F., and Martin, S. W. 2000. Relaxation in Glassforming Liquids and Amorphous Solids. *J. Appl. Phys.* 88 (6):3113 -3157.

Betancourt, Beatriz A. Pazmino, Douglas, Jack F., and Starr, Francis W. 2014. String Model for the Dynamics of Glass-forming Liquids. *J. Chem. Phys.* 140:204509.

Binder, Kurt, Baschnagel, Jörg, and Paul, Wolfgang. 2003. Glass Transition of Polymer Melts: Test of Theoretical Concepts by Computer Simulation. *Prog. Polym. Sci.* 28 (1):115–172.

Boyd, Richard H., Gee, Richard H., Han, Jie, and Jin, Yong. 1994. Conformational Dynamics in Bulk Polyethylene: A Molecular Dynamics Simulation Study. *J. Chem. Phys.* 101 (1):788-797.

Boyd, Richard, and Smith, Grant. 2007. *Polymer Dynamics and Relaxation*. New York: Cambridge University Press.

Brandrup, J., Immergut, E. H., and Grulke, E. A. 1999. *Polymer Handbook*. New York: Wiley-Interscience.

Brandrup, J., Immergut, E. H., and Grulke, E. A. 2003. *Polymer Handbook*. 4 ed. New York: Wiley-Interscience.

Brown, David, and Clarke, Julian H. R. 1990. A Direct Method of Studying Reaction Rates by Equilibrium Molecular Dynamics: Application to the Kinetics of Isomerization in Liquid n-butane. *J. Chem. Phys.* 92:3062-3073.

Byutner, Oleksiy G., and Smith, Grant D. 1999. Conformational Properties of Poly(vinylidene fluoride). A Quantum Chemistry Study of Model Compounds. *Macromolecules* 32 (25):8376-8382.

Byutner, Oleksiy G., and Smith, Grant D. 2000. Quantum Chemistry Based Force Field for Simulations of Poly(vinylidene fluoride). *Macromolecules* 33 (11):4264-4270.

Corezzi, Silvia, Fioretto, Daniele, and Rolla, Pierangelo. 2002. Bond-controlled Configurational Entropy Reduction in Chemical Vitrification. *Nature* 420:653-656.

Debenedetti, Pablo G., and Stillinger, Frank H. 2001. Supercooled Liquids and the Glass Transition. *Nature* 410:259-267.

DiMarzio, E. A., and Gibbs, J. H. 1958. Chain Stiffness and the Lattice Theory of Polymer Phases. *J. Chem. Phys.* 28:807-813.

Ediger, M. D. 2000. Spatially Heterogeneous Dynamics in Supercooled Liquids. *Annu. Rev. Phys. Chem.* 51:99-128.

Ediger, M. D., and Harrowell, Peter. 2012. Perspective: Supercooled Liquids and Glasses. *J. Chem. Phys.* 137:080901.

Ferrer, Maria Luisa, Lawrence, Christopher, Demirjian, Berj G., Kivelson, Daniel, Alba-Simionesco, Christiane, and Tarjus, Gilles. 1998. Supercooled Liquids and the Glass Transition: Temperature as the Control Variable. *J. Chem. Phys.* 109 (18):8010-8015.

Fox, Thomas G., and Flory, Paul J. 1950. Second-Order Transition Temperatures and Related Properties of Polystyrene. I. Influence of Molecular Weight. *J. Appl. Phys.* 21 (6):581-591.

Fukuda, Mitsuhiro, and Kikuchi, Hiroaki. 2000. Chain Dynamics and Conformational Transition in Cis-polyisoprene: Comparison between Melt and Subglass State by Molecular Dynamics Simulations. *J. Chem. Phys.* 113 (10):4433-4443.

Fulcher, Gordon S. 1925. Analysis of Recent Measurements of the Viscosity of Glasses. *J. Am. Ceram. Soc.* 8:339-355.

Gibbs, J. H., and DiMarzio, E. A. 1958. Nature of the Glass Transition and the Glassy State. *J. Chem. Phys.* 28:373-383.

Gotze, W., and Sjogren, L. 1992. Relaxation Processes in Supercooled Liquids. *Rep. Prog. Phys.* 55:241-376.

Han, Jie, Jaffe, Richard L., and Yoon, Do Y. 1997. Conformational Characteristics of Polymethylene Chains in Melts and in Various Phantom Chains from Explicit Atom Molecular Dynamics Simulations. *Macromolecules* 30 (23):7245-7252.

Hess, B., Kutzner, C., Spoel, D. van der, and Lindahl, E. 2008. GROMACS 4: Algorithms for Highly Efficient, Load-balanced, and Scalable Molecular Simulation. *J. Chem. Theory Comput.* 4:435-447.

Hess, Berk, Bekker, Henk, Berendsen, Herman J. C., and Fraaije, Johannes G. E. M. 1997. LINCS: A Linear Constraint Solver for Molecular Simulations. *J. Comput. Chem.* 18 (12):1463-1472.

Hoover, W. G. 1985. Canonical Dynamics: Equilibrium Phase-space Distributions. *Phys. Rev. A* 31 (3):1695-1697.

Kanaya, T., Kawaguchi, T., and Kaji, K. 1999. Local Dynamics of Some Bulk Polymers above T_g As Seen by Quasielastic Neutron Scattering. *Macromolecules* 32 (5):1672-1678.

Kanaya, Toshiji, Kaji, Keisuke, and Inoue, Kazuhiko. 1991. Local motions of cis-1,4-polybutadiene in the melt. A quasielastic neutron-scattering study. *Macromolecules* 24 (8):1826-1832.

Kauzmann, Walter. 1948. The Nature of the Glassy State and the Behavior of Liquids at Low Temperatures. *Chem. Rev.* 43 (2):219-256.

Kohlrausch, R. 1854. Theorie des elektrischen Rückstandes in der Leidner Flasche. *Ann. Phys. (Leipzig)* 91:56-82.

Liang, Taining, Yang, Yong, Guo, Dawei, and Yang, Xiaozhen. 2000. Conformational Transition Behavior Around Glass Transition Temperature. *J. Chem. Phys.* 112 (4):2016-2020.

Loncharich, Richard J., and Brooks, Bernard R. 1990. Temperature Dependence of Dynamics of Hydrated Myoglobin. *J. Mol. Biol.* 215:439-455.

Lyulin, Alexey V., and Michels, M. A. J. 2002. Molecular Dynamics Simulation of Bulk Atactic Polystyrene in the Vicinity of T_g. *Macromolecules* 35:1463-1472.

Nosé, Shumacrichi. 1984. A Molecular Dynamics Method for Simulations in the Canonical Ensemble. *Mol. Phys.* 52 (2):255-268.

Parrinello, M., and Rahman, A. 1981. Polymorphic Transitions in Single Crystals: A New Molecular Dynamics Method. *J. Appl. Phys.* 52 (12):7182-7190.

Peter, S., Napolitano, S., Meyer, H., Wubbenhorst, M., and Baschnagel, J. 2008. *Macromolecules* 41:7729-7743.

Qiu, XiaoHua, Moe, N. E., Ediger, M. D., and Fetters, Lewis J. 2000. Local and global dynamics of atactic polypropylene melts by multiple field ^{13}C nuclear magnetic resonance. *J. Chem. Phys.* 113 (7):2918-2926.

Schönhals, Andreas, Kremer, Friedrich, Hofmann, Achim, Fischer, Erhard W., and Schlosser, Eckard. 1993. Anomalies in the Scaling of the Dielectric α-relaxation. *Phys. Rev. Lett.* 70 (22):3459-3462.

Shavit, Amit, Douglas, Jack F., and Riggleman, Robert A. 2013. Evolution of Collective Motion in A Model Glass-forming Liquid During Physical Aging. *J. Chem. Phys.* 138:12A528.

Sillescu, Hans. 1999. Heterogeneity at the glass transition: a review. *J. Non-Cryst Solids* 243:81-108.

Singh, Sadanand, Ediger, M. D., and Pablo, Juan de. 2013. Ultrastable Glasses from in Silico Vapour Deposition. *Nature Mater.* 12:139-144.

Smith, Grant D., Borodin, Oleg, and Paul, Wolfgang. 2002. A Molecular-dynamics Simulation Study of Dielectric Relaxation in A 1,4-polybutadiene Melt. *J. Chem. Phys.* 117 (22):10350-10359.

Smith, Grant D., and Yoon, Do Y. 1994. Equilibrium and dynamic properties of polymethylene melts from molecular dynamics simulations. I. *n*-Tridecane. *J. Chem. Phys.* 100 (1):649-658.

Starr, Francis W., Douglas, Jack F., and Sastry, Srikanth. 2013. The Relationship of Dynamical Heterogeneity to the Adam-Gibbs and Random First-order Transition Theories of Glass Formation. *J. Chem. Phys.* 138:12A541.

Stukalin, Evgeny B., Douglas, Jack F., and Freed, Karl F. 2009. Application of the Entropy Theory of Glass Formation to Poly(a-olefins). *J. Chem. Phys.* 131:114905.

Takeuchi, Hisao, and Roe, Ryong-Joon. 1991. Molecular Dynamics Simulation of Local Chain Motion in Bulk Amorphous Polymers. II. Dynamics at Glass Transition. *J. Chem. Phys.* 94 (11):7458-7465.

Tammann, G., and Hesse, W. 1926. Die Abhängigkeit der Viscosität von der Temperatur bie unterkühlten Flüssigkeiten [The dependence of the viscosity on the temperature of supercooled liquids]. *Z. Anorg. Allg. Chem.* 156:245-257.

Tang, Qiyun, and Hu, Wenbing. 2014. Slowing Down of Accelerated Structural Relaxation in Ultrathin Polymer Films. *Phys. Rev. Lett.* 112:148306.

Vogel, H. 1921. Temperaturabhängigkeitsgesetz der Viskosität von Flüssigkeiten [Temperature dependence Law of the viscosity of liquids]. *Phys. Zeit.* 22:645-646.

Williams, G., Watts, D. C., Dev, S. B., and North, A. M. 1971. Further Considerations of Non-symmetrical Dielectric Relaxation Behaviour Arising from A Simple Empirical Decay Function. *Trans. Faraday Soc.* 67:1323-1335.

Williams, Graham, and Watts, David C. 1970. Non-symmetrical Dielectric Relaxation Behaviour Arising from A Simple Empirical Decay Function. *Trans. Faraday Soc.* 66:80-85.

Williams, Malcolm L., Landel, Robert F., and Ferry, John D. 1955. The Temperature Dependence of Relaxation Mechanisms in Amorphous Polymers and Other Glass-forming Liquids. *J. Am. Chem. Soc.* 77:3701-3707.

Xia, Wenjie, and Keten, Sinan. 2013. Coupled Effects of Substrate Adhesion and Intermolecular Forces on Polymer Thin Film Glass-Transition Behavior. *Langmuir* 29:12730-12736.

In: An Introduction to Glass Transition   ISBN: 978-1-53615-706-2
Editor: Roberta Ramirez   © 2019 Nova Science Publishers, Inc.

Chapter 3

A BRIEF HISTORY OF GLASS TRANSITION STUDIES IN WELL-DEFINED AMPHIPHILIC LIPID SYSTEMS

*Shigesaburo Ogawa**

Department of Materials and Life Science,
Seikei University, Tokyo, Japan

ABSTRACT

A glass transition is the phenomenon in which an amorphous phase exhibits abrupt changes in derivative thermodynamic properties. Many non-crystalline materials have the potential to exhibit the glass transition, while polymers, inorganic species, and sugars are excellent candidates to maintain a stable glass state under ambient conditions. Recently, investigations involving glass transitions in amphiphilic lipids (or surfactants) consisting of hydrophilic and lipophilic parts such as hydrocarbons have increased. This chapter provides a brief summary of the studies relevant to glass transitions in well-defined lipids systems such as anhydrous and/or water mixed systems. Then, some of the current problems and future tasks to be addressed in the 2020s will be described.

---

* Corresponding Author's E-mail: s.ogawa.chem@gmail.com.

Keywords: amphiphilic lipid, surfactant, glycolipid, phospholipid, glass
transition, glassy liquid crystal, lyotropic liquid crystal, gel

1. INTRODUCTION

In Section 1.1, generally accepted basic concepts of glass transitions are
initially described. Subsequently, "glass transition in liquid crystals" will be
introduced in Section 1.2. Fundamentally, "the glass transition of
amphiphilic lipid systems", which will be the main focus in this chapter, can
be scientifically included in the categories. Then, the purpose of this chapter
will be discussed in Section 1.3.

1.1. Glass Transition

An amorphous solid system is formed in a normal liquid if its
temperature is lowered rapidly enough from its melting temperature, T_m. The
process is referred to as a liquid-to-glass transition or glass transition
(Debenedetti and Stillinger 2001). In spite of the lack of periodicity of
crystals, the system's mechanical behavior is similar to that of a solid.

Significant changes in several thermophysical parameters of a liquid
occur within a relatively small temperature range around a characteristic
temperature, the so-called glass transition temperature (T_g) (Kauzmann
1948). In particular, the T_g detected by thermal analysis is called the
calorimetry T_g, and strictly speaking, is differentiated from the T_gs obtained
using other techniques. In brief, glassification drastically decreases the
specific heat and thermal expansion, similar to the crystallization
mechanism (Figure 1a & b). These changes are discontinuous functions of
temperature. On the other hand, parameters such as the specific volume and
heat content change continuously as functions of temperature, which differs
from crystallization, in which discontinuous changes occur in these
parameters (Figure 1c & d). Thus, the glassy state encompasses common
and different characteristics of both solid and liquid states. Considering that

glass is often defined as a "disordered (liquid-like) solid system" describes the system well. That is, the typical glassy state lacks long-range order and therefore can be considered an amorphous solid or non-crystalline solid. As a consequence, the glassy state possesses residual values in terms of entropy, heat content, and specific volume at 0 K (Kauzmann 1948). Owing to this metastable state, a slow relaxation process near T_g occurs, which is an important feature regarding the glass' stability (Wang et al. 2002). Typically, the dependence of the glass transition temperature T_g for a cooling rate Q obeys an exponential law:

$$Q = Q_0 exp\left[-\frac{E_g}{RT_g}\right] \tag{1}$$

where E_g is the activation energy of the structural relaxation process, R is the gas constant, and Q_0 is a constant. Here, the structural relaxation must take the form of enthalpy or volume. The calorimetric T_g is detected as a step-like phenomena, reflecting the heat capacity (or heat flow) change at T_g in a thermogram produced by differential scanning calorimetry (DSC) (Figure 2), and the relaxation degree of the glass can be estimated as the integrated area of the endothermic peak, which is called the excess enthalpy (enthalpy release) from non-annealed glass, in which structural relaxation has not occurred. Thus, for glass prepared under slow cooling rates, the endothermic peak occurs near T_g, reflecting enthalpy recovery after enthalpy relaxation, as observed in the heating DSC thermogram (Figure 2). The extent of relaxation, the integral of the endothermic peak, increases with decreasing cooling rate or with the annealing time below T_g.

The glass state is a thermodynamically metastable sate, while the crystal state is thermodynamically stable below T_m. However, because crystallization typically requires supercooling or supersaturation to enhance the probability of producing a new solid-liquid interface or solid-air interface (nucleation), rapid cooling can decrease the probability of crystallization by decreasing the time until the temperature reaches T_g and thus can lead to the glassy state. Using the proper cooling apparatus or methodology, successful glassification has been achieved for many materials, although some of these materials will readily crystallize if classical cooling methods are applied. Such findings regarding the basic criteria for glass-forming capabilities support the fundamental science behind glass physics. Glassification offers many diverse materials to use in compounds where crystalline solids cannot be applied. Therefore, understanding the glass forming capabilities is a fundamental issue and is expected to play a critical role in further developing useful applications in the near future.

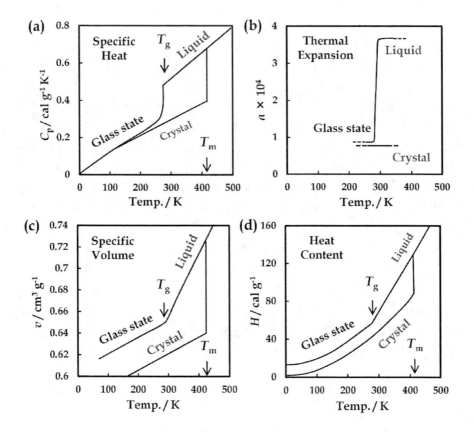

Figure 1. Thermodynamic properties of glucose in the liquid, crystal, and glassy states (Kauzmann 1948, p. 225).

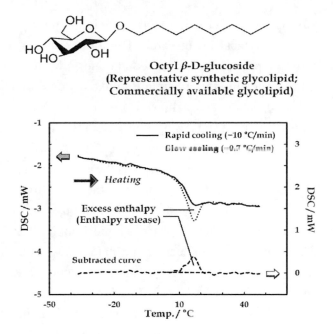

Octyl β-D-glucoside
(Representative synthetic glycolipid;
Commercially available glycolipid)

Figure 2. Heating DSC thermograms for anhydrous octyl β-D-glucoside cooled at different rates (− 0.7°C/min and − 10°C/min) prior to the measurement, shown along with the subtracted thermogram. A step-like phenomenon indicates the glass transition, and an increase in intensity of the endothermic-like peak or increase in the integrated area for the subtracted curve indicates an increase in the extent of excess enthalpy by slow cooling (Ogawa 2016).

1.2. Glass Transition in Liquid Crystals

Liquid crystals (LCs) are materials that include both fluidity and periodicity (long-range ordering), meaning that LCs share the anisotropic properties of crystalline materials and the fluid properties of an isotropic liquid (IL). Therefore, they are regarded as intermediate phases between crystalline solids and amorphous fluid liquids, or so-called "mesophases". Importantly, the LC state is obtained as a thermodynamically stable phase, and the transformation from or to other phases requires a change in latent heat at the transition temperature. There are various ways to classify LC materials depending on molecular features and the type of molecular assembly (the type of LC phase), and on how the LC phases have been

Shigesaburo Ogawa

obtained. Regarding the last term, the thermotropic LC phases or lyotropic LC phases are used in typical LC research. The former phases are obtained by the effect of temperature on pure compounds or on a mixture of compounds, either by heating the crystalline solid or by cooling in an IL. On the other hand, the latter are achieved by mixing compounds with appropriate solvents over a range of concentrations and temperatures. Typically, compounds which can form the LC phase exhibit a double melting behavior during the heating process as the crystal phase is heated to the IL state (Figure 3); The transformation from the crystalline phase to the LC phase occurs, and subsequently, the transformation from the LC phase to the IL state occurs. Here, if the LC phase is rapidly cooled below T_m while avoiding crystallization, the molecular arrangements with long-range ordering or anisotropy can be frozen in the solid state, resulting in the formation of glassy liquid crystals (GLCs) (Figure 3). Typical step-like phenomena are distinctly observed at calorimetric T_g in both the cooling and heating processes (Figure 4). The phase transition from LC to IL in the heating process in the higher temperature regions provides solid evidence that the glass transition occurred for the LC phase.

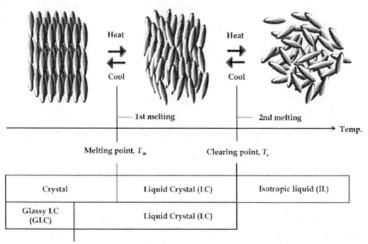

Figure 3. Double melting behavior referring to T_m and T_c and the glassification of the LC phase under supercooling conditions at temperatures below T_g for LC-forming materials.

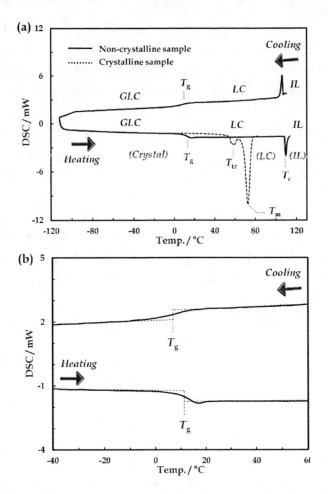

Figure 4. (a) Full and (b) enlarged views of the DSC thermograms of the anhydrous octyl β-D-glucoside. The phase states and the defined phenomena observed for both non-crystalline and crystalline samples are depicted around the thermogram in (a). The phase states given in parentheses are for the crystalline sample. T_{tr} indicates the temperature at which a crystal-to-crystal phase transition occurs for anhydrous octyl β-D-glucoside occurs (Ogawa et al. 2013). In subfigure (b), the determination of the middle T_gs during both cooling and heating is described; the T_gs are determined as the temperatures corresponding to half of the magnitude of ΔC_p or the heat-flow change at T_g (Ogawa et al. 2010; Ogawa and Osanai 2012).

Approximately 50 years have passed since the pioneering discovery of the glassy state of the LC phase. In 1968, Sackmann reported on the possible formation of a GLC state for an LC "mixture" of cholesteryl chloride and cholesteryl myristate, but no description in terms of the thermal aspects of the glass transition was provided (Sackmann et al. 1968; Sackmann 1968). On the other hand, the accurate observation of a calorimetry glass transition of LC consisting of a "pure" low molecular weight compound such as cholesteryl hydrogen phthalate was reported by Tsuji et al. (1971), and by Sorai and Seki (1971). Since then, proper studies aimed at demonstrating the thermal aspects inherent in GLC have been published and the formation of GLC in nematic and smectic LC phases consisting of a pure compound was observed (Sorai et al. 1984; Sorai 1985). These studies include the discovery of multi-step glass transitions for a single-compound system, and its layered structure was considered as the characteristic feature that differentiated GLC from typical glassy materials. The term multi-step (or double glass) transition will be discussed later in this chapter. Currently, GLC materials are attracting significant attention in the development of optoelectric and conductive devices (Chen et al. 1999; Percec et al. 2002; Chen et al. 2014; Kobayashi et al. 2017).

However, while GLC materials consisting of hydrophobic compounds have been largely developed, GLC materials formed by amphiphilic compounds such as amphiphilic lipids have been studied to a lesser extent. Diverse amphiphilic lipids exist and are applied in aqua-related systems, which are predominantly associated with the life sciences. Therefore, understanding glassy materials consisting of amphiphilic lipids should be particularly important in that sense, and differentiating them from GLC materials consisting of hydrophobic compounds should prove beneficial.

1.3. Purpose of this Chapter

An enormous variety of amphiphilic lipid compounds are present in nature and have been synthetically prepared to date. Amphiphilic lipid molecules consist of both hydrophilic and lipophilic (hydrophobic) parts. Their typical molecules can actively interact through a variety of interfaces (e.g., air/liquid, liquid/liquid, liquid/solid, and air/solid) and among themselves (Figure 5). They readily form monolayer films at air/liquid interfaces (so-called Langmuir film or Gibbs adsorption film (Eastoe and Dalton 2000; Israelachvili 1994; Langmuir 1916; Langmuir 1917), and can form emulsified systems (Bancroft 1913; Tiddy 1980; Winsor 1968), which result from the reduction of surface tension through adsorption at liquid/liquid interfaces (wetting at the interface). In addition, even without an interface, these amphiphilic lipids can self-assemble into colloidal assemblies such as liposomes (vesicles), micelle aggregates, or lyotropic LC phases in aqueous solutions or water-mixed systems (Kunitake 1992; Tiddy 1980; Winsor 1968).

The hydrophilic part includes a polar region, which exhibits strong interactions (e.g., hydrogen bonding or ionic interactions), which are significant driving forces in the development of intra- and inter- molecular interactions. However, for typical lipophilic domains (e.g., hydrocarbon chains), only weak van der Waals interactions, or so-called London dispersion forces, exist, which are not significant forces. For typical amphiphilic lipid molecules that unsymmetrically contain both polar and nonpolar regions, the microsegregation structure (where the individual hydrophilic parts develop their inter-molecular interactions without any distinct interaction between hydrophilic and hydrophobic parts), are energetically favorable. As a consequence, the LC phase of lipids consisting of a bilayer structure is frequently formed because of the microsegregated structure (Kunitake 1992; Tiddy 1980; Winsor 1968). With variations of concentration or temperature, various types of lyotropic LC phases, such as fluid lamellar (L_α), hexagonal columnar, or cubic phases occur. It is worth noting that these assemblies can solubilize or miscible with the polar or nonpolar species in their lyophilic or lipophilic domains. In addition, in the low temperature regions of the phase diagram for the amphiphilic lipids and water systems, not only the crystal phase, but also the lamellar gel phase (L_β), in which the hydrocarbon tail becomes solid-like while the hydrophilic part remains disordered (Kunitake 1992; Tiddy 1980; Winsor 1968), are often present (Figure 5, below). The fluidity of the LC phase drastically changes in the gel phase. The gel phase is sometimes abbreviated "g" in the phase diagram, but it does not relate in any way to the "glassy state". The transition to or from the gel phase is accompanied by latent heat; therefore it is categorized as a first order transition, meaning that the gel is categorized as one type of phase. In general, there are many types of phases possible in amphiphilic lipid-water systems (Figure 5).

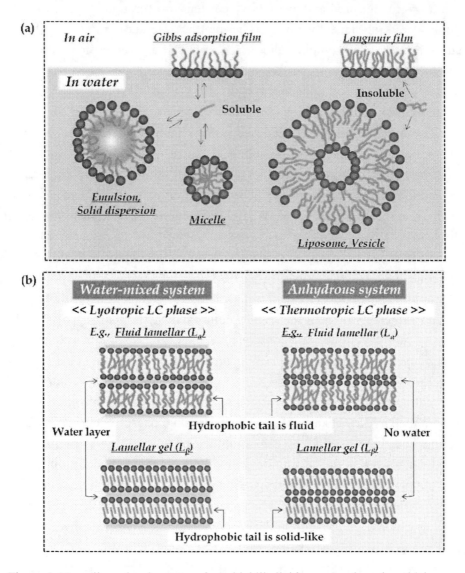

Figure 5. Versatile molecular states of amphiphilic lipid compounds such as (a) in water systems and (b) LC systems. In both lamellar phases (L_α and L_β), the hydrophilic part is assumed as disordered, regardless of the state of hydrophobic part.

On the other hand, the glassy state is not a type of phase. That is, the glass transition occurs in a "defined" phase such as an IL (solution or melt), an LC, or a gel phase. Therefore, it is inevitable that the study of the glass transition in lipids has been performed as part of a phase study or by its challengeable extension, where the phase studies themselves are highly complicated. In addition, the great difficulty in detecting an actual glass transition in colloidal systems (owing to the complexity of these systems) must undoubtedly impede accurate investigation of the glass transition in these systems. However, focus on the glass transition behavior of these systems has been growing recently.

Practically, T_g has been considered as the most important single parameter that should be known in order to evaluate the potential applications of many materials. Regarding aqua-related applications, for example, T_g can determine the processability, product properties, quality, stability, and safety of food systems (Slade et al. 1991). The importance of constructing the phase diagram incorporating equilibrium and non-equilibrium parameters such as nucleation temperature, devitrification temperature, recrystallization temperature, and T_g curves is well-understood (MacKenzie 1977). The findings from such phase diagrams aid in the identification of the kinetic and thermodynamic factors responsible for the cryopreservation of living cells; typically, the glassy matrix can restrict the mobility of molecules and the diffusion of active chemicals, effectively inhibiting the aggregation, gelation and deterioration of the biological molecules. With these points in mind, we believe that understanding the basic behavior of glass transitions in amphiphilic lipids systems is valuable toward further applications of these materials. A summarization of previous studies will aid in conveniently understanding the current knowledge in this research field. In this chapter, a brief historical summary of glass transition studies in amphiphilic lipid systems, in particular, of well-defined systems consisting of "pure" or "water mixed" amphiphilic lipids whose hydrophobic part is a hydrocarbon chain, is presented.

2. GLASS TRANSITION STUDY OF WELL-DEFINED AMPHIPHILIC LIPIDS SYSTEMS

2.1. Glass Transition Studies of Amphiphilic Lipids Systems (1980s)

In the 1980s, researchers began to note the possibility of the occurrence of the glass transition in amphiphilic lipid systems. Initial investigations on glass transitions in pure lipids or lipid/water systems began from various aspects, depending on the backgrounds of the researchers.

Syûzo Seki, who was an Eminent Professor of Osaka University, Japan, was one of the first outstanding researchers in the field of glassy science. Seki's research group, consisting of various outstanding researchers, has studied various types of glassy materials including inorganic and organic compounds, and ice using high performance calorimetry (e.g., adiabatic calorimetry).

Kodama and Seki studied the thermal behaviors of ionic surfactant dioctadecyldimethylammonium chloride (DOAC)/water systems. During their research, Kodama and Seki discovered the glass transition in the DOAC – water system for the supercooled complete gel phase (Kodama et al. 1981). The occurrence of the glass transition was recognized at approximately − 40°C as step-like phenomena in the DSC thermogram. From this observation, they proposed the "glassy gel phase" descriptor. The phenomenon was observed only when a large amount of water (with 51 g% of DOAC) was mixed, and was not recognized in anhydrous DOAC.

John W. Goodby, emeritus professor of York University, USA, is one of the outstanding researchers in the field of LC materials. A variety of novel GLC materials have been developed by his research group; this chapter only focuses on the glass transition of selected amphiphilic lipid systems. In an article published in 1984, Goodby commented that alkyl β-D-glucoside transitions to a glassy thermotropic smectic A phase solid (Goodby 1984). He considered the formation to be a "glassy smectic LC phase" based on the observation of a slight rise in the baseline of the cooling thermogram of DSC and in the solid-like mechanical features of the sample studied; however, the thermogram and quantitative data were not presented in the article.

While the above two research groups commented on the glass transitions of synthetic lipids, Blöcher and Ring et al. (1984) reported on the "glass transition in biological lipids" extracted from *Thermoplasma acidophilum* (MPL). The glass transitions of lipids were observed for the glycophospholipid fraction, the main component of the phospholipid fraction, and the glycolipid fraction. Subsequently, they reported on the glass transition of "pure" diphytanylglucosylglycerol compounds consisting of synthetically prepared glycolipid and extracted lipids mainly consisting of phospholipids (Blöcher et al. 1985). A clear glass transition behavior was recognized in a thermogram obtained by differential thermal analysis (DTA) for the synthesized glycolipid. However, it must be mentioned that they observed the glass transition in the presence of a buffer (i.e., 400 mM sodium cacodylate (MW: 160.0)-HCl (pH 7.0)) and ethylene glycol (MW: 62.7) (12.5 M)), meaning that there were impurities present in the lipids systems. Investigations of phase transitions including glass-transition phenomena for MPL and buffer mixtures were continued by Blöcher and Ring et al. (1990) and Ernst et al. (1998). In a report related to the impurity problem, Melchior et al. (1982) had detected the glass transition for cardiolipin solution by DSC analysis after quick freezing; however, Hirsh (1983) demonstrated that the glass transition was virtually identical to the glass transition observed in $CaCl_2$, which was used for the preparation of calcium cardiolipin.

Excluding the above selected articles, other reports mentioned the possibility of glass transitions. For instance, Marcus and Finn (1985) reported that a mixture of decyl glucoside and dodecyl glucoside (possibly β-anomers) at a one-to-one ratio formed a glassy smectic LC phase at room temperature. However, they did not provide experimental evidence. In addition, because the T_g values of the individual "pure" compounds were below 13°C (Ogawa et al. 2013), the validity of the results could be questionable. However, if mixing decyl and dodecyl glucosides can actually greatly enhance their individual T_gs values, then this would be a very valuable observation. Gliozzi et al. (1986) discussed the possibility of the occurrence of glass transition-like phenomena for the extracted lipids from the plasma membrane of *Sulfolobus solfataricus* for a fixed quantity of water at 2.8% without the addition of buffer species. They included the thermogram as evidence, but the step-like phenomenon in the thermogram coincided with other transition peaks, meaning that further verification would be required to provide conclusive results. Although Costello and Gulik-Krzywicki (1976) had, on the basis of X-ray methods, proposed the formation of a condensed "glass-like" packing for liquid paraffin chains after freezing, Melchior (1986) reported that no indication of glass transition was detected by DSC for many different types of lipid bilayers of various lipid composition as well as for various native membrane preparations.

Thus, although significant preliminary observations were reported, the concept of "glass transition in amphiphilic lipid systems" was not fully confirmed in the 1980s because of the lack of experimental evidence in the majority of the published articles.

2.2. Glass Transition Studies of Amphiphilic Lipids Systems (1990s)

In the 1990s, several novel insights were reported. For instance, in terms of the glycolipid species, Auzély-Velty and Goodby et al. reported the occurrence of glass transition in the thermotropic LC phase consisting of synthetic diether-type archaeal glycolipid compounds possessing one furanosyl unit below − 25°C (Auzély-Velty et al. 1998b). In addition, the occurrence of a glass transition for the thermotropic LC phase of tetraether-type archaeal glycolipid compounds possessing two furanosyl units was reported (Auzély-Velty et al. 1998a). In the latter article, the two T_gs values of the compounds were listed in Table 1, but no appropriate explanation was provided. DSC thermograms were not shown in the articles, and therefore the corresponding quantitative values were unknown. However, from these articles, the formation of a "glassy hexagonal columnar LC phase" was clearly described.

Also, the formation of a "glassy bicontinuous cubic LC phase" was possibly observed in a study by Fischer and Vill et al. (1994). In the article, they obtained a DSC thermogram for a 1,1-di-(β-D-glucosyl-oxymethyl)pentadecane, Y-shape type glycolipid compound possessing two sugar heads, exhibiting distinct step-like phenomena in both cooling and heating cycles. These phenomena indicated a glass transition. The authors did not discuss the phenomena as a glass transition, but because the glass transition appeared to take place in the cubic LC phase, the formation of a "glassy cubic LC phase" with the *Ia3d* space group appears to have taken place. Later, Minden et al. (2000) reported the formation of a glassy cubic LC phase for octadecyl α-D-lactoside as well as a glassy smectic LC phase for oleoyl β-D-maltoside, although no evidential results were provided (not even T_g) and only a textual description was provided. The formation of colorless glass for a glycolipid of the 2/2 type (two sugar and two lipophilic chains) was described by Schmidt and Jankowski (1996).

While Blöcher et al. had reported on the occurrence of glass transitions for extracted lipids, which somewhat included phospholipids and buffer components, the glass transition derived from "pure" phospholipids had not yet been specifically mentioned. Voinova (1995) theoretically treated a bilayer made from dimyristoyl phosphatidylcholine (DMPC) or dipalmitoyl phosphatidylcholine (DPPC), as a fluid surface which undergoes cooperative structural transitions from an initial LC state to a quasi-solid gel state or glass-like state using slow or fast cooling, respectively. When the rate of cooling reaches a critical value, the formation of large spherical solid domains is halted, and they named this phenomenon as a "membrane glass-like transition" or "surface vitrification". That is, they proposed a surface membrane glass-like transition for pure phospholipids. A theoretical analysis was also presented shortly thereafter (Voinova 1996). At least, Voinova considered that the formation of the gel phase and glassy state occur separately, whereas the formation of a "glassy gel phase" resulting from the glass transition in the gel phase was previously described by Kodama and Seki (1981) for an ionic surfactant system.

Thus, new studies regarding the glass transition of amphiphilic lipid systems emerged in the 1990s. However, the coherency among groups was limited.

2.3. Glass Transition Studies of Amphiphilic Lipids Systems (2000 to Present)

From 2000 on, detailed systematic descriptions, together with clear evidence of glass transitions and associated behaviors, have been published. As a result, the structural dependencies of glass transitions in amphiphilic lipids systems have been somewhat clarified. In this section, the contents are divided into three parts: 2.3.1. Glycolipids Systems, 2.3.2. Phospholipids Systems, and 2.3.3. Other Lipids Systems. In addition, because investigations on the glass transition in "pure" and "water-mixed" glycolipids systems have systematically increased, the section on glycolipids is further divided into two parts: 2.3.1.1. Glycolipids Systems (2000 to 2009) and 2.3.1.2. Glycolipids Systems (2010 to present).

2.3.1. Glass Transition Studies of Glycolipids Systems from 2000

A glycolipid consists of a sugar moiety and a hydrophobic tail. There are various types of glycolipids owing to various types of sugar and hydrophobic structures. Significant efforts have been made to investigate the formation of an LC phase for glycolipid compounds (Goodby et al. 2007; Hashim et al. 2018). Also, the formation of the glassy LC phase for various glycolipid compounds has been observed, although a limited number of reports provide detailed descriptions. Dumoulin and Goodby et al. stated that "The lack of formation of glassy states is unusual for such liquid crystalline glycolipids, which often tend to exhibit a liquid crystal to glassy transition upon cooling with the retention of the structure of the liquid crystal phase" (Dumoulin et al. 2002, p. 13744). However, it was noted that some glycolipids readily recrystallize during the cooling process even at relatively high cooling rates.

In the 2000s, the glass transition phenomena for typical alkyl glycosides consisting of a single tail and single sugar head with one, two, or three sugar moieties were reported in detail (Ericsson et al. 2005a; Ericsson et al. 2005b; Hoffmann et al. 2000; Kocherbitov and Söderman 2004; Ogawa et al. 2010; Ogawa et al. 2013; Singh et al. 2010), as well as for alkyl glycosides with methyl-branched alkyl chain (Milkereit et al. 2005b), Guerbet branched chain glycosides with double-chain tail (Ahmad et al. 2012; Patrick et al. 2018; Saari et al. 2018), Y-shaped glycolipids with a variety of double head alkyl glycosides consisting of double sugar heads and a single tail or double tail (Jana et al. 2006; Jayaraman et al. 2007; Milkereit et al. 2004; Milkereit et al. 2005a; Milkereit and Garamus 2005), ethyl-branched neoglycolipids (Milkereit and Garamus 2005), sugar esters such as sucrose ester, trehalose ester, lactose ester, and raffinose ester (Molinier et al 2006; Ogawa et al. 2016a; Ogawa et al. 2018; Perinelli et al. 2018; Szűts et al. 2007) and etc (Chambert et al. 2007; Singh et al. 2013).

2.3.1.1. Glass Transition Studies of Glycolipids Systems (2000 to 2009)

Some systematic investigations were conducted by Hoffmann et al. (2000). They reported the glass transition behavior of a thermotropic LC phase formed by alkyl α-D-glucosides with a chain number between 7 and 10. In their report, a clear step-like phenomenon in the baseline of the DSC thermogram was present. They indicated that the thermotropic phases can be strongly supercooled such that glass transitions at T_g are observed. Importantly, they proved that T_g/T_ms were 0.79 ±0.01, regardless of the chain length. Such a relationship between T_g and T_m had been proven for other glassy materials, but none included these particular compounds. Thus, coherent reports on such thermophysical parameters including T_g for low molecular amphiphilic lipids were originated by Hoffmann et al. However, for the many studies pertaining to glycolipid research that were included, the glass transition was not the main topic and no "glass" term appeared in the Abstract or Keywords of the article.

On the other hand, Kocherbitov and Söderman focused on the glass transition phenomenon as the main topic rather than as incidental to a phase behavior study (Kocherbitov and Söderman 2004). They focused on the undefined solid forms of alkyl maltosides at that time and reported a "glassy fluid lamellar (L_α) LC phase" formed by octyl β-D-maltoside and decyl β-D-maltoside, and reported on the effects of traces of water on the state. By using a DSC, small-angle X-ray scattering (SAXS), and water sorption calorimetry, they proved that the glassy LC phases turn into lyotropic Lα LC phases upon sorption of water at constant temperature, and thus the plasticizing effect on T_g was also demonstrated.

Later in 2005, Ericsson and Ulvenlund et al. (including Kocherbitov and Söderman as co-authors) demonstrated the occurrence of glass transitions of tetradecyl β-D-maltoside and hexadecyl β-D-maltoside and discussed the distinctive influence of kinetics on the phase behaviors, including the glass transition. While the glassy state of octyl β-D-maltoside and decyl β-D-maltoside must be regarded as "frozen" versions of a fluid lamellar LC phase, they proposed a "glassy lamellar gel (L_β) phase" for tetradecyl β-D-maltoside and hexadecyl β-D-maltoside, in which the alkyl chains are ordered in a hexagonal array and hydrophilic parts are disordered glass, respectively. In the same year, Ericsson and Ulvenlund et al. further investigated the effect of the number of sugar moieties on the glass transition (2005b). They observed the glass transition of dodecyl β-D-maltoside and dodecyl β-D-maltotrioside in DSC thermograms. Because the T_gs increases from 65°C for dodecyl β-D-maltoside (a disaccharide derivative) to 100°C for dodecyl β-D-maltotrioside (a trisaccharide derivative) because the head-group length increases by one glucose unit, they thus assumed that the increase requires more thermal energy to transform the glass to the LC phase when the number of intermolecular hydrogen bonds between head-groups increases. They considered that the glass transition temperature of dodecyl β-D-maltoside and dodecyl β-D-maltotrioside is primarily governed by the nature of their head-groups. However, they could not observe the glass transition of dodecyl β-D-glucoside (monosaccharide derivative) after lyophilization.

In the same timeframe, Milkereit, and Vill et al. reported on the glass transition of thermotropic LC phases for inverted Y-shaped glyco-glycero-lipids, N-palmitoyl-1,3-di-O-β-D-glucopyranosyl-2-deoxy-2-amino-sn-glycerol (Milkereit et al. 2004), and of siamese-twin shaped glycolipids (Milkereit et al. 2005a). In the former case, the formation of a glassy columnar hexagonal LC phase or glassy cubic LC phase was suggested. In the latter case, the formation of a glassy smectic LC phase was suggested. However, no DSC thermogram or T_g was presented in either study. Jana et al. also mentioned the glass transition to a glassy columnar hexagonal LC phase for double-headed glycolipids without the determination of T_g. It was noted that "the temperature at which the mesophase freezed into glassy state could not be ascertained" (Jana et al. 2006, p. 534), suggesting that the accurate determination of T_gs was difficult for these compounds. On the other hand, Jayaraman et al. (2007) successfully reported the T_g values of analog glycolipids. In their report, the material was hydrated and therefore the second heating scan became informative as the hydrates evaporated at approximately 100°C. They reported the effect of the alkyl chain length between 8 to 16 on T_g. Most exhibited a glass transition at approximately 60°C, but some compounds with an alkyl chain length of 8 or 10 were irregularly reduced. They speculated that "the presence of two glucose units in the molecule might lead to strong hydrophilic head group interactions and probably hinders free movement of the molecules with respect to each other" (Jayaraman et al. 2007, p. 2229).

On the other hand, Milkereit and Garamus and Milkereit and Vill et al. studied the glass transition occurring for alkyl maltosides with ethyl-branched neoglycolipid (Milkereit and Garamus 2005) and methyl-branched maltoside (Milkereit et al. 2005b). They indicated the presence of the glassy state using optical polarising microscopy, but thermograms showing the step-like phenomena were not provided. They considered proof of the glass formation as the appearance of cracks resulting from shrinking during cooling (Milkereit et al. 2005b). However, ambiguous explanations were presented in some of these articles. For instance, they mentioned that "On cooling from the isotropic phase the SmA phase formed and at room temperature it took several hours until the glass was formed again" (Milkereit and Garamus 2005, p. 157). However, glassification must occur quickly if the temperature falls below T_g. Another example is "Figure 5 shows a typical example of the DSC thermogram of the Mel-β-Phy/water system (49 wt%), in which an endothermic peak (ΔH) 38 kJ/mol) associated with the glass $\leftrightarrow$ Lα phase transition of Mel-β-Phy was observed between 33 and 43°C." (Milkereit et al. 2005b). The sentence suggests that one might mistake the glassy state as a kind of phase.

In 2006, the glass transition of sugar ester was clearly demonstrated with both cooling and heating thermograms by Molinier et al. (2006).

Synthetically prepared 1'-*O*-sucrose mono ester exhibited the glass transition at approximately 46.2°C in both cooling and heating processes. The reverse step-like phenomena observed in the DSC thermogram at similar temperatures clearly demonstrated the reversibility of the glass transition.

Somewhat later, Szüts et al. (2007) studied the thermal behavior of commercial sucrose monoesters (mixtures including isomers and some di-derivatives). They used a modulated-temperature DSC (MDSC), which allows the relaxation endotherm and glass transition to be separated into non-reversing and reversing signals, respectively. They demonstrated the existence of the glass transition, although the step-like phenomenon coincides with the melting behavior of the materials in the thermogram. The authors also proposed the applicability of "hot-melt technology" to sugar esters, because they vitrify when melted.

Thus, during this time period, the features of the glassy state for glycolipids were further identified and demonstrated, and provided future study topics for these materials. In some review articles, the glass transition was recognized, although the corresponding discussions lacked full content (Claesson et al. 2006; Goodby et al. 2007; Singh and Jayaraman 2009).

2.3.1.2. Glass Transition Studies of Glycolipids Systems (2010 to Present)

In a report by Kocherbitov and Söderman (2004, p. 3058) it was mentioned that "the glass transition and the glassy state were not observed in the studies of octyl α-D-glucoside and octyl β-D-glucoside". From this observation, Kocherbitov and Söderman concluded (p. 3058), "in the case of maltosides, that the bulky headgroups interact with each other strongly than glucoside headgroups do and therefore can more easily form the glassy state". The study by Ericsson and Ulvenlund (2005b) further supported that the glassy state was not observed in dodecyl β-D-glucoside. Such a concept developed from these experimental observations was treated as significant in later research and was cited in a review article (Claesson et al. 2006). However, as mentioned above, Hoffmann et al. had clearly shown the occurrence of the glass transition of alkyl α-D-glucosides in 2000. In addition, the glass transition of alkyl β-D-glucoside was clearly reported in 2010 (Ogawa et al. 2010). The thermograms are shown in Figure 2 and Figure 4 of this chapter.

From 2007 to the present, the current authors have published several studies related to the glass transition in pure glycolipids, water-mixed and water and ice coexistence systems. In 2010, we recognized and demonstrated the glass transition of octyl β-D-glucoside and octyl β-D-thioglucoside and their water systems (Ogawa et al. 2010). The glassification of the smectic phase in the absence of water and of the lyotropic lamellar and lyotropic cubic LC phases of those compounds in the presence of water were observed. In the paper, the glass transition behavior in terms of "lyotropic glassy LC phases" were clearly described. Anomalous relationships regarding the heat capacity change ΔC_p at T_g for glycolipids and water mixture system with glycolipids mole fraction were observed for both octyl β-D-glucoside and octyl β-D-thioglucoside-water systems. To interpret their behavior, we adopted the concept of "non-continuous water" to propose a hypothesis (Ogawa et al. 2010; Ogawa and Osanai 2012). In addition, the fitting of T_g-concentration curves using an empirical model described by Couchman-Karasz equations was carried out. Couchman-Karasz equations are written in the form of Equation (2) (Couchman and Karasz 1978; Couchman 1978), in which the glass transition phenomenon can be treated as a second order phase transition. The detailed derivation of the equations has been provided elsewhere (Miller et al. 1998; Pinal 2008). The general Couchman-Karasz formula for glycolipid-water binary systems is

$$lnT_g = \frac{x_1 lnTg_1 + kx_2 lnT_{g2}}{x_1 + kx_2} \tag{2}$$

where the subscripts 1 and 2 denote the components: a glycolipid and water, respectively. The symbols x_1 and x_2 represent the mole fractions of the corresponding components. T_g is the glass transition temperature of the mixture under consideration; k is a constant defined as $\Delta C_{p2}/\Delta C_{p1}$, where ΔC_{p1} and ΔC_{p2} are the ΔC_p at T_g of pure component 1 (anhydrous glycolipid) and component 2 (pure water), respectively. If the T_g of a glycolipid-water mixture of uncertain concentration is measured, the water content of the system can be roughly estimated using Equation (2).

In 2012, the current authors reported on the glassification in the freezing-thawing processes of octyl β-D-glucoside-water systems. The concentration-temperature phase diagram of the octyl β-D-glucoside-water binary system, which included the melting temperatures of ice, lyotropic phase transition temperatures, devitrification temperatures, and glass transition temperatures, T_g and T_g', of the unfrozen phases in the absence and presence of ice (Ogawa et al. 2012). Ice nucleation temperatures and devitrification temperatures were provided in a previous report (Ogawa et al. 2010) and were summarized in a recent review chapter (Ogawa 2016). The phase diagram is depicted in Figure 6 of this chapter.

In the 2012 report, the phase diagram and analysis by cryo-polarizing optical microscopy demonstrated the inhibitory effects of octyl β-D-glucoside on the nucleation and growth of ice crystals in extremely high concentration systems, in which the formation of a "glassy Lα LC phase in the presence of ice" was anticipated. Alternatively, the formation of a "glassy micelle state" was also proposed for a pentyl β-D-glucoside-water system (Ogawa 2016).

Such glassification properties of glycolipids, which greatly depend on the sugar moiety number in the molecule, are significant in preventing the precipitation of the eutectic composed of electrolytes and ice during freezing-thawing processes (Ogawa and Osanai 2012). The occurrence of eutectic is typically responsible for damage to cells and enzymes and eutectic is considered to induce active species as a result of unusual pH changes. From these results, we believe that glycolipids compounds exhibiting high glass-forming capabilities can be excellent candidates to protect against protein activity during freeze-drying processes.

Actually, Izutsu et al. have engaged in such pioneering work, and have shown that when a small amount of sucrose ester is added to a system, it exhibits excellent effects in the maintanance of protein activities during freeze-thawing and freeze-drying processes (Izutsu et al. 1993; Izutsu et al. 1994; Izutsu et al. 1995). More recently, Imamura et al. (2014) mentioned a similar property for a greater variety of protein targets. However, no relationship with the glass-forming capabilities of glycolipids was discussed, but Izutsu et al. (1995) noted that the amorphous state can be effectively persisted when a sucrose ester is added as the stabilizer.

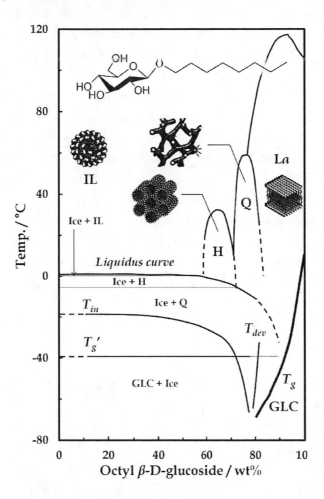

Figure 6. Global phase diagram of octyl β-D-glucoside-water system containing the nonequilibrium parameters such as ice nucleation curve (T_{in}), devitrification curve (T_{dev}) and glass transition temperature curves (T_g and T_g') (Ogawa et al. 2010; Ogawa et al. 2012; Ogawa 2016). The abbreviations of H and Q indicate the hexagonal and cubic LC phases, respectively.

We considered that the glassification of the freeze-dried matrix may be the answer to why sucrose ester is an excellent lyoprotectant. With this in mind, we investigated the effect of glycolipids on protein activity during both freeze-thawing and freeze-drying processes, focusing on the relationship between the glass forming ability and the stabilization effect. As a consequence, we have shown that an increase in the alkyl chain length in the glycolipid molecules exerts positive effects on the maintenance effect of protein activity during the freeze-thawing process (Ogawa et al. 2015). On the other hand, bifunctional properties such as an amphiphilic nature (a surface activity) and glass forming capabilities for glycolipid molecules are crucial for effective stabilization using a low excipient concentration of glycolipids during the freeze-drying process and subsequent storage at room temperature (Ogawa et al. 2016). Glycolipids consisting of oligosaccharide with a high T_g promote the retention of enzymatic activity, whereas the glycolipid consisting of a monosaccharide with a low T_g performed poorly at high concentrations, albeit much better than Tween 80 with a much lower T_g at medium concentrations. That is, the applicability of the bifunctional property of glycolipids compounds exhibiting a glass-forming capability under dry situations as well as surface activity in an aqueous solution to act as stabilizers for proteins was proposed. The concept of the bifunctional property had been considered as a future research direction in a previous paper (Ogawa et al. 2013).

On the other hand, the occurrence of the glass transition for natural-like branched-chain glycosides such as 2-hexyldecyl β and α-D-maltoside mixtures was suggested based on their heating and cooling DSC thermograms (Ahmad et al. 2012). More recently, Patrick and Hashim et al. (2018) and Saari, Hashim, and Zahid et al. (2018) have reported on the glass transition behavior of the thermotropic LC phase, and on the behavior of the lyotropic LC phase, respectively, by changing the alkyl chain length. It was proposed that the self-assembly of both dry and fully hydrated Guerbet maltosides, which exhibited both glass-forming capabilities and surface activity, and could also act as membrane-stabilizing compounds, makes them ideal candidates for practical use in the food industry as well as in biomedical research.

Meanwhile, Kocherbitov and Söderman made important remarks in 2004 that "the hydrocarbon part becomes glassy in the same transition as the headgroup part. Nevertheless, since the transition to the glassy state of the whole system is caused by the decrease of mobility of the headgroup layer, the interactions between the hydrocarbon chains play a minor role in this particular glass transition. On the contrary, the hydrocarbon part cannot keep the liquid structure when the surrounding headgroup layers making up most of the mass of the system stop moving" (2004, p 3059).

However, Nowacka and Topgaard et al. (2013) reported the effects of hydration and temperature on the dynamics of octyl β-D-maltoside by using an 13C MAS NMR analysis. The data for the PT ssNMR glass transition temperature was determined by analogy with the definition of the 2H NMR glass transition temperature. They demonstrated that the hydrophilic headgroup of the compound is a "PT ssNMR glass", while the hydrophobic tail is still a "PT ssNMR liquid" at the calorimetric glass transition. Thus, they concluded that "characterizing the glass-transition in a surfactant system with a single glass-transition temperature fails to account for the wide range of dynamics occurring on the molecular scale" (p. 174), which is opposite to the expectations of Kocherbiotov and Söderman (2004). It should be noted that the PT ssNMR glass transition temperature for the hydrophilic part reported by Nowacka and Topagaard et al. was 70°C higher than the calorimetric temperature reported by Kocherbitov and Söderman; therefore, further verification is needed to close this issue.

In a similar timeframe, the current authors reported the effect of the alkyl chain length on the glass transition behavior of alkyl β-D-glucosides with the alkyl chain length ranging from 1 – 12 (Ogawa et al. 2013; Ogawa 2016). This research was intrinsically conducted to clarify the differences in the glass transition between non-amphiphilic sugars and LC-forming amphiphilic sugars (glycolipids), because the alkyl β-D-glucoside with short chain length cannot form the LC phase and therefore, glassification occurs in the IL state. The plasticization tendency of the alkyl β-D-glucosides was clearly observed when the alkyl chain length changed from 1 to 5. This is responsible for the reduction in the hydrogen bonding interaction induced by the increase in hydrophobic volume. However, as the alkyl chain length reaches the length which induces the formation of the LC phase of alkyl β-D-glucosides, the decrease in T_g stops and T_g becomes constant, regardless of the chain length. In addition, no clear increase in ΔC_p [J/mol] was discernible, meaning that no contribution by the alkyl chain part was recognized. Based on these results and the results obtained by powder X-ray diffraction analysis (PXRD) for the supercooled species, we concluded that the strength of the hydrogen bonding between sugar moieties may be the dominant factor in the glassification of LC forming glucosides. However, with regard to the hydrophobic part, two possibilities exist: either their end-side interdigitated hydrocarbon chain affects their glassification very weakly or it does not affect it at all. A decisive conclusion was not obtained. Along with these series of investigations, "chain-length dependent enthalpy relaxation tendency" was also studied for alkyl β-D-glucosides with the alkyl chain length ranging from 1–12 (Ogawa 2016).

More recently, the current authors evaluated the glass transition of sugar esters such as trehalose and raffinose esters (Ogawa et al. 2016). High T_g values (80 – 87°C) were noted for these lipids, while the T_g values of sucrose esters were below 50°C. In 2018, we studied the effects of the acyl chain length on the glass transition behavior of trehalose lipids, as well as the dehydration and thermotropic LC phase behaviors (Ogawa et al. 2018). Similar to the case of alkyl β-D-glucosides, no distinct effects due to differences in the hydrocarbon chain length were observed regarding T_g and ΔC_p in case the alkyl chain length is enough to form the LC phase. Moreover, the effects of the phase structure, such as the Lα or Q phase, on T_g and ΔC_p were compared and it was confirmed that the type of LC phase structure did not directly affect these values. Also, dehydration-induced metastable LC phase behavior was clearly observed, and surprisingly, we discovered the occurrence of a reversible phase transition between Lα and Lβ phases, even below the calorimetric T_g (Figure 7). Thus, the "phase transitions between glassy Lα and glassy Lβ phases" were identified. These results, and in particular the reversibility, strongly support the idea that some part of the alkyl chain domain must remain fluid.

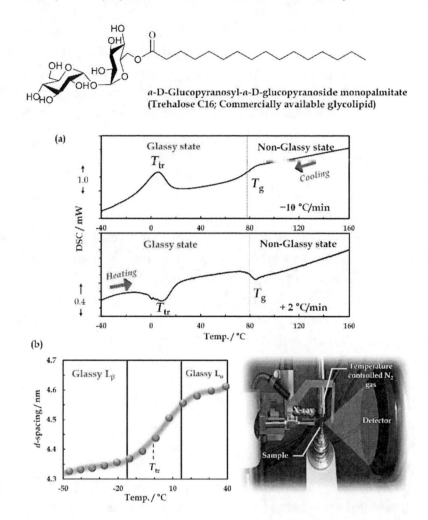

a-D-Glucopyranosyl-*a*-D-glucopyranoside monopalmitate
(Trehalose C16; Commercially available glycolipid)

Figure 7. (a) DSC thermograms and (b) temperature-dependent wide-angle X-ray
scattering (WAXS) profiles for anhydrous Trehalose C16 glycolipid and the
experimental setup for low-temperature WAXS analysis using a Saturn 70 CCD
system. Apparently, a transition accompanied by large latent heat occurred reversibly
under the glassy state in both cooling and heating processes (Figure 7a). This behavior
was explained by the phase transition between the L_α and L_β phases, which is
accompanied by a distinct variation of the packing structure of hydrocarbon chains.
The temperature-dependence of the WAXS results obtained with the experimental
setup shown in (b) provides solid evidence supporting this hypothesis
(Ogawa et al. 2018).

As the phase transition from LC to gel phases is simply explained by the solidification of the alkyl chain part, the glassification of the sugar moiety would not prevent the transition if the alkyl chain domain were not in the glassy state. However, the complicated gel-LC phase transition process for lipid systems has not been sufficiently defined by themselves, and further studies in this area are required.

During this time period, the original concept of the glassy state as applied to amphiphilic glycolipids has been significantly amplified and emphasized. At the same time, experiments on the use of the glass-forming capabilities of glycolipids continues to advance. An article that reviewed the topic of the glass transition of glycolipids was published by Hashim and Ogawa et al. (2018), and a book chapter that mainly featured the glass transition behavior under anhydrous and water systems considering both the absence and presence of ice for alkyl glucosides was published (Ogawa 2016; Ogawa and Osanai 2012). However, based on the results related to glassy science for glycolipids systems, there remain specific unsolved issues, and additional research is needed.

2.3.2. Glass Transition Studies of Phospholipids Systems
(from 2000 to Present)

An enormous variety of studies pertaining to biologically occurring phospholipid compounds have been reported to date because of the importance of these compounds in various applications. The glass transition in well-modeled phospholipid systems, in particular, has attracted extensive attention regarding the stabilization of phospholipid membranes.

As it has been reported that the accumulation of some solutes such as sugars is associated with the capability to survive either freezing or desiccation in a diverse array of organisms, the effect of sugars on the stability of biological membranes has attracted significant attention. Under this concept, numerous investigations on the lyotropic phase behavior of phospholipids in the presence of sugars have been conducted. Earlier, Crowe et al. (1998) proposed the "water replacement hypothesis" which addresses the direct interaction of sugar molecules with the polar groups of phospholipids. On the other hand, Koster et al. (2000) confirmed that a depression of the melting point, T_m, is caused by the rigidity of the sugar glass rather than by a direct lipid-sugar interaction.

With these results as background, Shalaev et al. (2001) began to study the effect of sugar on the phase behavior of 1,2-dioleoylphosphatidylethanolamine (DOPE). Interestingly, they observed two T_g events in the heating thermogram of DOPE-sugar mixtures dehydrated over phosphorus pentoxide at room temperature. They concluded from a comparison of the types of sugar additives that the glass transition at the lower temperature should be attributed to the glass transition associated with the DOPE phase rather than the sugar phase. Subsequently, they studied a synthetic phospholipid, 1-palmitoyl-2-oleoyl-*sn*-glycero-3-phosphoethanol amine (POPE) in the absence of sugar additives regarding the glass transition behavior and its nonexponential relaxation (enthalpy relaxation) behavior below T_g and the plasticization caused by water (Shalaev and Steponkus 2003). They observed the formation of a "glassy L_β phase for phospholipid-water systems" such as a POPE-water mixture. At the same time, they claimed that the glass transition in phospholipids is possibly associated with the rotational rather than the translational motion. More recently, they systematically studied the glass transitions of several phosphatidylcholine (PC) and phosphoethanol amine (PE) species in the presence of water (Shalaev and Steponkus, 2010). Glass transitions occur for several phospholipid-water systems made using DMPC, DPPC, 1,2-dioleoyl-*sn*-glycero-3-phosphocholine (DOPC), DOPE, and POPE. The Gordon-Taylor equation has been used to describe the plastization effects of water on the T_g. The Gordon-Taylor equation (Gordon et al. 1977) is written as follows;

$$T_g = \frac{x_1 T g_1 + k x_2 T_{g2}}{x_1 + k x_2} \tag{3}$$

where T_{g1} and T_{g2} are the glass transition temperatures of the anhydrous phospholipid and water, respectively, and k is a constant. The Gordon-Taylor equation is an empirical equation and a similar form can be introduced by modification of the Couchman-Karasz equation if T_{g1}/T_{g2} is not significantly different from unity (Couchman 1978) or if the heating capacity change is proportional to the temperature (Brinke et al. 1983). The problematic issue when using the modified equation was well described by Katkov and Levine (2004). For glycolipid-water systems, the application of Equation (2) is more recommended than Equation (3) (Ogawa et al. 2010).

According to a report by Shalaev and Steponkus (2010), the majority of lipids have k values of 6.4–8.8, whereas lower values were obtained for DOPE; it is noteworthy that T_{gl} and k were fitted parameters and T_{gl} was not determined experimentally. The L_β phase of DOPE has a k value of 3.5, whereas the two non-lamellar phases, the ripple phase and the inverted hexagonal phase, possess even lower k values of 1.3 and 1.4. The authors further showed the effects of T_m and T_g. The T_g/T_m ratio increases with dehydration in linear fashion, with a similar slope for different lipids. T_g/T_m varies between 0.7 to above 0.9 for DMPC at low water contents.

Although the possibility of the glass transition in phospholipids like DMPC was proposed by Voinova (1995), a systematic investigation including conventional calorimetric analysis by Shalaev and Steponkus et al. (2003, 2010) indicates that one must successfully generalize the glass transition of synthetic phospholipid-water systems, which can be expected to be a common feature of biological membranes and phospholipids bilayer preparations such as liposomes. Shalaev et al. (2016) recently published a review article, where the glass transition studies of phospholipids included these topics to a certain extent.

In addition to the reports by Shalaev et al. and Vinova, the glass transition behavior or glass-like state of synthetic phospholipid-water systems have been addressed in various studies by using pressure perturbation calorimetry measurements (Wang and Epand 2004), electron paramagnetic resonance (EPR) spectroscopy and/or neutron scattering (Erilov et al. 2004; Isaev and Dzuba 2008; Aloi and Bartucci 2017), anelastic spectroscopy (Castellano et al. 2006), quasielastic neutron scattering (Swenson et al. 2008), broadband dielectric spectroscopy together with MDSC (Svanberg et al. 2009; Berntsen et al. 2011), and others. However, Swenson et al. mentioned that "the gel-to-liquid transition seems to have the same impact on the lipid dynamics as the glass transition has for ordinary glass forming liquids" (Swenson et al. 2008, 045101-6), suggesting that a distinct clarification of the relationship between the gel-to-liquid transition and glass transition is required for adequate understanding of the process. In addition, Shalaev and Steponkus (2010) have posed a question concerning what groups are involved in such T_g-associated relationships (e.g., headgroup or hydrocarbon tail) and have pointed out that the glass transition of "confined water" in the phospholipid mesophase has not been clarified. That is, phospholipids glass appear to exhibit still complicated behaviors. Nevertheless, scientific interest regarding the glass transition behavior of synthetic phospholipids has been growing and further progress is expected.

2.3.3. Glass Transition Studies on Other Amphiphilic Lipids or Surfactant Systems (from 2000 to Present)

Glass transition behaviors have been observed not only for amphiphilic lipids such as glycolipids and phospholipids but also other amphiphiles and surfactant systems. Although the applicability for their glass-forming capabilities is unknown because their T_g values are typically much lower than $0°C$, scientific observations and a professional investigation have recently been performed.

As mentioned previously, Kodama and Seki (1981) studied kinetic-dependent behaviors in gel phase formation and the formation of the glassy state in the gel phase. On the other hand, Jensen et al. (1998) demonstrated the occurrence of the glass transition for Triton X-100 at $- 59°C$, where the surfactant acts as a plasticizer on the miscible epoxy system. Later, Júnior and Petri et al. (2009) and Amim Jr. and Petri et al. (2012) reported that the glass transition for surfactants such as polyoxyethylenesorbitan monolaurate (Tween 20), polyoxyethylenesorbitan monopalmitate (Tween 40), and polyoxyethylenesorbitan monostearate (Tween 60) occurs at approximately $- 61°C$ and the addition of these surfactants induces the plasticization of polymers such as carboxymethylcellulose acetate butyrate and cellulose esters. In their reports, the glass transition for Span 40 could not be obtained owing to its high crystallization tendency, but the plasticization effect was recognized in the polymer mixed system.

Meanwhile, since the second half of 2000s, the research group led by Professor Calum J. Drummond has reported on numerous glass-forming low-molecular amphiphilic lipids and on their versatile lyotropic phase behavior. The hydrophobic tails of these lipids are unsaturated tails or isoprenoid branched tails, which effectively reduce the crystallization tendency compared to normal saturated hydrocarbon tails.

Gong and Drummond et al. (2008) reported the glass transition of urea derivatives such as 3,7,11,15-tetramethylhexadecyl urea and 3,7,11-trimethyldodecyl urea, at approximately − 80°C and also reported on the glass transition of the 5'-deoxy-5-fluoro-N^4-(phytanyloxycarbonyl) cytidine amphiphilic lipid (2011). Notably, they reported the "double glass transition phenomena" and stated that "it may contain two amorphous domains; a disordered (glassy) hydrophobic chain and a disordered (glassy) headgroup region, and that the two domains undergo separate glass transitions" (2011, p. 1554).

Fong and Drummond et al. reported on the glass transition for phytanyl ethylene oxide (EO) surfactants. They confirmed that a series of enthylene oxide surfactants equipped with a 3,7,11,15-tetramethyldodecyl chain and 3,7,11-trimethyldodecyl chain exhibited glass transitions. The T_gs gradually increased from − 90.0°C to − 72.1°C as the number of EO units increased up to 7 for EO derivatives equipped with a 3,7,11,15-tetramethyldodecyl chain (Fong et al. 2010). On the other hand, the T_gs for EO surfactant equipped with a 3,7,11-trimethyldodecyl increased as the number of EO units increased up to 5, from − 98.9°C to − 79.5°C but was constant when the number of EO units increased from 5 to 8 (Fong et al. 2011).

In addition, Sagnella and Drummond et al. reported the glass transition of ethanolamide derivatives such as monoethanolamide with isoprenoid-type hydrocarbon chains (Sagnella et al. 2009) and diethanolamide with isoprenoid-type hydrocarbon chains (Sagnella et al. 2011a) or unsaturated C18 hydrocarbon chains (Sagnella et al. 2011b), and endocannabinoid analogues of anandamide such as 2-arachidonyl glycerol, arachidonyl dopamine, 2-arachidonyl glycerol ether (noladin ether), and o-arachidonyl ethanolamide (virodhamine) compounds (Sagnella et al. 2011c), respectively. T_gs for monoethanolamide and diethanoamide derivatives were observed around −73 °C, regardless of the difference in hydrophobic structures between 3,7,11-trimethyldodecyl and 3,7,11,15-

tetramethylhexadecyl chains (Sagnella et al. 2011a; Sagnella et al. 2011b). On the other hand, the difference in degree of unsaturation for diethanolamide derivatives affect their T_gs. As the degree of unsaturation increases from one (Oleoyl) to three (γ-Linolenoyl) for C18 hydrocarbon chain, the decrease in T_g was reported (Sagnella et al. 2011c). It was noted that two glass transitions in the heating thermogram were suggested for some of endocannabinoid analogue compounds, such as 2-arachidonyl glycerol, 2-arachidonyl glycerol ether and virodhamine. Similar to the report by Gong et al. (2011), Sagnella expected that the lower T_g was attributed to the glass transition in the hydrocarbon layer and the higher one was attributed to that in the hydrophilic headgroup, respectively (Sagnella et al. 2011c). Also, Moghaddam and Drummond et al. reported the occurrence of two glass-like transitions for the ethylene diamine tetraacetic acid bis phytanyl compound, whereas a single glass transition was reported for the corresponding mono phytanyl derivative (2010).

Thus, Drummond et al. reported the numerous glass-forming amphiphiles with unsaturated tails or isoprenoid branched chains, which effectively reduce the crystallization tendency more than the normal saturated hydrocarbon tail, and successfully induce glassification. As these glasses were formed in the thermotropic LC phase, these compounds must form glassy LC phases. Most of the compounds show low T_g and none of the investigation on their glass transition behavior were carried out in the presence of water. In addition, the verification for the assignment of the glass transition was only determined in the single DSC thermogram, and the confirmation from the enthalpy relaxation phenomenon had not been carried out; glass transition has not been focused mainly in those reports.

On the other hand, Merino et al. evaluated the influence of thermal treatment on the glass transition behavior of Triton X-100 in both the bulk (Merino et al. 2011) and confined states in nanoporous inorganic materials (Merino et al. 2013b; Merino et al. 2013a). In their works, not only the DSC analysis but also the dielectric relaxation spectroscopy were used for the investigations. As mentioned, the glass-forming property of Triton X-100 had already been reported (Jensen et al. 1998), and, in the report (Merino et al. 2011), Merino et al. determined the "lower cooling limit to make glassy state" without the crystallization. Under their experimental conditions, a fully amorphous material was obtained when cooling from the liquid state at cooling rates higher than 10°C/min. In addition, the typical step-like jump at the glass transition in the DSC thermogram and the real permittivity in the dielectric relaxation spectroscopy measurement were recognized. It was noted that the fragility index for Triton X-100 was also successfully estimated. The "fragility index m" is a quantitative measure which expresses the degree of deviation from an Arrhenius-type temperature dependence near the T_g, and the fragility index, m, is defined as the derivative as follows (Angel 2002; Wang et al. 2006):

$$m = -\left(\frac{d \log \tau(T)}{d(T_g/T)}\right)_{T=T_g} \tag{4}$$

where τ is the temperature dependence of the average relaxation time; the deviation from exponential decay in the linear response to perturbations from the equilibrium state, and the nonlinearity of that response to thermodynamic perturbations. The fragility index is received as an important parameter to classify glass forming liquid materials. In the report by Merino et al., the devitrification occurs near T_g, and they proposed that one of the reasons must arise from a high dynamically fragile behavior. Merino et al. later studied the effect of crystallization on the detected relaxation processes by the dielectric relaxation spectroscopy (Merino et al. 2013b). In addition, Merino clearly showed the glass transition behavior of Triton X-100 included in a silica-based nanostructured matrix of 5.7 nm in pore diameter; T_g increased ~ 20°C (Merino et al. 2013a). They observed the effective suppression of crystallization by inclusion, together with the MDSC and enthalpy relaxation studies. That is, the "glass transition behavior of low-molecular surfactants with low T_g" was clearly described in these articles, whereas the effect of water was not studied.

Meanwhile, Häckl et al. reported the glass transition around at -25°C for carnitine alkyl ester bromides with alkyl chain length with 4 and 6 (Häckl et al. 2017). The formation of the LC phase was not studied. It was shown that the aqueous solution of these compounds affords the distinct critical aggregation concentration (CAC) or the critical micelle concentration (CMC) from the surface tension analysis. Therefore, glassification in some aggregation states was strongly expected.

Although the glass transition of these lipids introduced in this section were typically observed at much lower temperature regions, and the water systems were rarely studied, several distinguishing discoveries for the glass transition study in amphiphilic lipids or surfactant systems have been reported.

3. COMMENTS ON PROBLEMATIC ISSUES AND FUTURE TASKS

3.1. Problematic Issues

3.1.1. Problematic Issue on Experimental Notes

Since the glass transition accompanies abrupt changes not only in derivative thermodynamic properties but also in mechanical properties, the phenomenon can be noted from the various measurements or observations. The lack of crystallization in the deep supercooling state must increase the possibility of this circumstance. However, until relatively recently, precise and detailed descriptions have been few in terms of the glass transition in amphiphilic lipids systems, despite the GLC materials consisting of totally hydrophobic materials that have been greatly studied. Based on my experience in the glass transition study of mono-tailed glycolipids, I can say that convincing that precise investigation of the glass transition was not an easy task.

Amphiphilic lipids are readily adsorbed with traces of water in very short times. I have experienced that even if an anhydrous sample was obtained by complete desiccation with heating and P_2O_5 desiccant under reduced pressures, in case that the sample was located under the ambient atmosphere for a short time, the reduction of T_g by plasticization with adsorbing trace water was inevitable (Ogawa 2016). This is because of the amphiphilic nature of the lipids, where fully hydrophobic compounds would not have the same water sorption issues. Even longer adsorption times sometimes yields a step-like behavior in the DSC thermograms into broad, obscured, and double steps depending on the mixing state of adsorbed water. Thermal annealing at high temperature, which can homogenize the sample liquid or LC state, can help to clarify the thermogram (one step glass transition can be obtained, even though the reduction of T_g was inevitable). Although the features of amphiphilic lipids are the fundamental issues for these studies, there are less papers that refer to them. Thermogravimetric analysis for glycolipids shows the gradual reduction of sample mass below 100°C, if trace water is adsorbed (Ishak et al.).

Kocherbitov and Söderman (2003) and Ericsson et al. (2005b) mentioned that if trace amounts of water are present, the phase behavior of glycolipids can be changed. For instance, they clearly showed the sample with trace amounts of water may coincide with an irregular pre-transition phenomenon attributed to the melting of hydrated crystals in the heating thermogram. According to their reports, the phase behavior of hydrate glycolipids was reported as an anhydrous sample. Ericsson et al. mentioned that "in particular, the credibility of data interpretation hinges on careful consideration of the question whether a given experiment is conducted under conditions where the water is retained during measurement, or not" (Ericsson et al. 2005b, p. 1535). Though they did not mention it as related to the issue of glass transition, the trace amount of water also greatly influenced the glass transition behavior (Kocherbitov and Söderman 2004; Ogawa et al. 2010; Ogawa 2016). It is believed that the detailed investigations on the glass transition of mono-tailed glycolipids by Kocherbitov and Söderman (2004) and Ericsson et al. (2005a; 2005b) must be successfully carried out as the results of their serious considerations about these issues.

When first studied, the glass transition of alkyl β-D-glucosides with a variety of alkyl chain lengths from 1 to 12 (Ogawa et al. 2013), the T_g for butyl β-D-glucoside was irregularly observed at much lower temperatures compared to other glucosides (not reported elsewhere). However, later it was found that the compound readily formed hydrated crystals comprising 4 moles of glycosides per mole of water, and the T_g of the micellar-like melt was greatly reduced by the water plasticization. Such hydrated crystal formations might be relatively random for the glycolipids, but strong hydrated crystal formation capability must sometimes impede the accurate investigation in the anhydrous state.

Thus, I would like to emphasize the problem of the adsorption of trace water must always be taken into consideration when the anhydrous state is discussed. On the other hand, complete desiccation for phospholipids systems might be impossible, and, until now, no report on the glass transition in the anhydrous form had been reported, whereas the water content dependence of phospholipid properties under desiccated conditions are greatly attractive research topics.

3.1.2. Problematic Issue on Data Interpretations

The lack of reproducibility must unavoidably occur if the trace of water is not taken consideration, in particular, if a sealed container is used for the DSC measurement. Conversely, the measurement using an open container can avoid the effect of water by pouring inert gas to evaporate water completely by thermal annealing (Jayaraman et al. 2007; Ogawa 2016). It is unknown whether the lack of quantitative descriptions in terms of glass transition of amphiphilic lipids had come from the lack of reproducibility. On the other hand, it is true that the description on glass transition has not been supported by "solid" experimental evidence for amphiphilic lipid systems, at least until the beginning of 2000s. During this period, the DSC thermogram showing a clearly step-like phenomenon for the anhydrous sample were rarely shown (many referencees), or the step-like phenomenon had not been interpreted as the glass transition (Boyd et al. 2000; Fischer et al. 1994). It is believed that the glass transition phenomena of these compounds has not been well understood for the researchers treating related compounds until relatively recently.

Although the glass transition is one of the choices to consider when an anomalous thermal event is recognized in the thermogram. When the step-like phenomenon was not clearly observed and other phenomena, like melting behavior or phase transition between LC phases, coincides, great care should be taken to assign it properly. For instance, the glass transition seemed to be observed around $-52°C$ for Phyt 1-(2-hydroxyethyl)-1-alkyl urea (Fong et al. 2007). However, seeing the DSC thermogram, the phenomenon as described as T_g is rather an endothermic peak. That is, it is probable that the thermal event cannot be readily considered as the sign of glass transition. Alternatively, the occurrence of the glass transition was reported for stearyl glucosides by Ishak et al. (2019). According to their report, stearyl α-D-glucoside exhibited a glass transition around 65°C, while it was observed around 35°C for stearyl β-D-glucoside. But, they mentioned that the glass transition occurred in the presence of a crystal, since after the "phenomenon", which they attributed as glass transition, clear melting occurred. If true, it means that the glass transition occurs in the crystalline state of stearyl α-D-glucoside. Actually, there are such materials which exhibit glass transitions in typically defined crystal states (e.g., hexagonal ice) (Suga and Seki 1980; Shalaev 2016). The step-like behavior that they mentioned as the glass transition can be also regarded as endothermic behavior, which may be assigned as the pre-melting or crystal-to-crystal transition, often observed for glycolipid crystals. For the stearyl β-D-glucoside, clear step-like phenomena was not recognized in the cooling process around the T_g observed in the heating thermogram. If the typical glass transition exists, reversible phenomena occur around a similar temperature in both the heating and cooling processes (Figure 4). Since there is not enough information, my assumption is, of course, not the conclusion. However, I think when facing a complicated thermogram, the assignment of the glass transition should be carefully carried out. The confirmation of enthalpy relaxation (Merino et al. 2013b; Ogawa 2016; Shalaev and Steponkus 2001; Shalaev and Steponkus 2003; Shalaev et al. 2010) or MDSC (Ericsson et al. 2005a; Ericsson et al. 2005b; Merino et al. 2013a; Szűts et al. 2007) analysis must be considered as a follow up characterization. The estimation and comparison of heat capacity change, ΔC_p, must be also effective (Hoffmann et al. 2000; Kocherbitov and Söderman 2004; Merino et al. 2013; Ogawa et al. 2013; Ogawa et al. 2018).

On the other hand, some reports have mentioned the presence of glass transition by using optical microscopy observations. Gerber et al. reported the phase behavior study of monoacylated amide-linked disaccharide glycolipids without the use of thermal analysis, such as DSC (2009). High T_g (above 110°C) was observed for the disaccharide-based glycolipids, and it was comparable to the T_m of analogue compounds. Though the microscopic observation might be an important tool in understanding the overall physical state of the matrix, support via thermal analysis should also be utilized. Clemente et al. reported the melting for monoacylated amide-linked disaccharide glycolipids with C16 fatty acid ester around 145°C in the DSC thermogram (2011), though Gerber et al. had mentioned the occurrence of glass transition around the same temperature (2009). T_g determination via microscopic observation might be an intrinsically different parameter from calorimetric T_g. Furthermore, Saari et al. showed 70°C in the Table 1 of their report for T_g of dodecyl β-D-maltoside (2018). The value seemed to be cited from the report by Koeltzow and Urfer (1984), though the value was not shown in the report. However, the description showed that there may be a relationship between the calorimetric T_g and softening point. The softening point is the point which is classically assumed as the point at which the van der Waals forces between the alkyl chains are disrupted (Koeltzow and Urfer 1984). However, while the calorimetric T_g is reported around 65°C (Ericsson et al. 2005b), the softening point was described around 87 − 112°C, showing poor correlation. Molinier et al. described a certain solid state as "soft solid," distinguished from the glassy state (2006). The clarification of any relationship between the "glassy solid" and "soft solid" states of these compounds will be an intriguing research topic.

3.2. Future Tasks

3.2.1. Systematic and Quantitative Investigations on the Relationship between Molecular Structure and Glass Forming Ability

The effect of branched hydrocarbon chains on the tendency of the glass-forming ability for glycolipids was suggested by Blöcher et al. (1985). The group of Professor Drummond has shown many examples, such as the introduction of branched and unsaturated alkyl chains to allow supercooling in compounds. However, a quantitative analysis has not yet been reported; the glass transition has not been the main focus in the aforementioned reports. It must be important how such functional substitutes quantitatively affect the glass forming ability. As Merino et al. (2011) have shown in the example of Triton X-100, the determination of the cooling limit and detailed investigations on the devitrification phenomenon can allow for better understanding of how the compound should be modified to improve and control the behavior under supercooling. Getting such insight would directly contribute to broader applications. For instance, the crystallization of amphiphilic lipid molecules can affect (typically reduce) the stabilizing effects on therapeutic proteins during freeze-thawing and freeze-drying processes. Therefore, successful control of the crystallization behavior via those processes have strong influence on the final product. Ultra rapid heating can be adopted to avoid devitrification (or cold crystallization).

At the same time, detailed structural analyses of the glassy state is desirable. In addition to thermal analysis, structural analysis to accurately define the glassy state (e.g., glassy LC phase, glassy micelle state) is important for not only for the application but also for the accurate understanding of the relationship between the type of molecular assembly and the thermophysical characteristics.

As mentioned in Section 1.2., Sorai et al. have proposed that the multi-step glass transition for single compound systems can be the characteristic feature of layered structures (1985). It is important to confirm whether the concept agrees with the case for amphiphilic lipid systems. An XRD-DSC technique is a powerful tool to attain information relevant to both thermal and structural properties (Ogawa and Osanai 2012). However, the introduction of the cooling unit to the X-ray apparatus is not typical. Instead, since the experimental setting for single crystal X-ray structural analysis is typically equipped with a high performance cooling apparatus, it can be applied for the investigations on the glass transition in anhydrous LC phases (Figure 7b) (Ogawa et al. 2018). As introduced by Nowacka and Topgaard et al. (2013), the defining physical state (glassy or liquid) by individual atoms of the molecule using 13C MAS NMR analysis can be a great methodology if the experimental validity is further developed or generalized enough to estimate the calorimetric T_g by this technique. The introduction of different sample states such as aligned films might be a useful approach to gain deeper understanding of the structure of their glassy state.

In the case of amphiphilic lipid systems, the inhomogeneous phase containing highly concentrated or diluted portions in water systems or non-crystalline portions and a partial solidified gel portion in an anhydrous state are considered to coexist. In such a case, it might also exhibit the glass transition separately at different temperatures in individual phases. However, in addition to the thermal analysis, the introduction of further systematic and quantitative investigations, together with the introduction and development of structural analysis, will clarify this issue.

3.2.2. Investigations on Glass Transition Behavior in Water Systems

The ability of carbohydrates to form glassy matrices is a well-known phenomenon that has great importance in the fields of cryobiology, microbiology, room-temperature stabilization of biological materials for pharmaceutics, and food preservatives. In addition, the introduction of amphiphilic lipid (surfactant) has successfully constructed the nano-sugar glass systems, where the glassy sugar matrix was encapsulated by hydrophilic moiety of the amphiphilic lipid (Giri et al. 2011). Here, it is believed that if the lipid possesses glass-forming ability, at least the hydrophilic part must form the glassy domain in the GLC state even without the addition of other glass forming materials. That is, amphiphilic lipids equipped with surface activities in water systems and LC-forming and glass-forming abilities in water-mixed and under desiccated situations can simultaneously construct the self-organized molecularly assembled glassy system by freeze-drying of the corresponding amphiphilic lipid solution. This would be an innovative to prepare novel drug delivery materials.

However, these processes need the understanding of the phase diagram including not only the parameters under thermodynamic equilibrium, such as melting point and LC-LC phase boundaries, but also parameters such as T_g curve, T_g' curve, ice nucleation temperature and devitrification temperature curves obtained under far from equilibrium conditions as shown in Figure 6. Unfortunately, such studies have been rarely carried out so far except in a few works (Ogawa et al. 2012; Ogawa 2016). Glass transition studies of well-defined amphiphilic lipids in the presence of water is rare (Kocherbitov and Södermann 2004; Ogawa et al. 2010; Ogawa et al. 2012; Ogawa and Osanai 2012; Ogawa 2016; Shalaev and Steponkus 2003; Shalaev et al. 2010). The study of glass transition behavior for water systems, as well as anhydrous systems, must be important for the exploration of novel applications as well as for the understanding of the phase behavior of amphiphilic lipids-water systems. In the latter, it is believed that better insight into the glass transition behavior of the water system can induce the novel insight in the phase behavior study. In the report (Ogawa et al. 2010; Ogawa and Osanai 2012), the discontinuous state of water in the extremely concentrated (desiccated) system was proposed for the lyotropic LC phase as the background of the anomalous relationship between the ΔC_p at T_g and the glycolipid mole fraction. Such thermophysical parameters can become key parameters to enhance unknown phase behaviors that have not been studied from such aspects.

3.2.3. Contributions to the Development of General Glassy Science

Some articles have mentioned the possibility of the unique glassy state for amphiphilic lipids systems, in which the hydrocarbon chain can be in a liquid-like state, based on experimental evidence. Although the evidence needs further verification, if it is true, the glassy state after the primary glass transition has occurred consists of glassy hydrophilic domain and liquid-like hydrophobic domains. Such material must exhibit sophisticated functions since hydrophilic regions can contribute to the effective preservation of biomolecules and drugs, and fluid hydrophobic regions can be miscible to oily media or solubilize the hydrophobic species, allowing the material to be an excellent candidate for fascinating drug-delivery systems or novel molecular systems.

On the other hand, imaging the system, in the case of lamellar, hexagonal or cubic LC phases form glassy states, where the hydrocarbon region is still in the liquid state. With this information, it is possible to determine whether such systems can be attractive to consider the ideal glassy state system (like free-standing polymer film), since the aggregative properties by the intermolecular interactions will greatly influence the glass transition behavior. Although Anderson claimed that the theory of the glass and glass transition is the "deepest and most interesting unsolved problem in solid state theory" (Anderson 1995, p. 1615), the study on fundamental glassy science, using such dimensionalized self-assembled glassy systems might for the simplification of the target system and provide a substantial intellectual spin-off.

CONCLUSION

Many studies have noted the glass transition behavior in well-defined amphiphilic lipids systems and have cultivated various notions regarding the glass transition of amphiphilic lipids systems to date, e.g., glassy gel phase, glassy thermotropic LC phases (such as smectic, hexagonally columnar, cubic LC phases), glassy lyotropic LC phases (such lamellar and cubic), gel-fluid lamellar transitions in the glassy state, kinetic effects, water plasticization, etc. In addition, some potent applications have been proposed (such as a stabilizer for biomaterials during freeze-drying and storage and as a host material for hot-melt technology). Thus, the glass transition behavior in-well defined amphiphilic lipids systems should not be regarded as a rare phenomenon. Nevertheless, systematic studies have been few, because most of the work was performed as part of a study on phase behavior. Taking into consideration that the use of glassy materials can be tremendously beneficial in the current industry, specific focus on the glass transition behavior of these systems should prove of significant value.

Along with the further cultivation of the consensus pertaining to the problematic issue common to studies of the glass transition in amphiphilic lipid systems and the verification based on solid experimental evidence, further clarification of the specific behavior of these systems is essential. In particular, the elucidation and control of the state of the hydrocarbon tail is an urgent issue to be resolved and should contribute to the development of novel application fields for both these materials and glassy science, in general. In this regard, complementary comparisons using different analytical techniques would be of significant benefit; for instance, the unification of the calorimetric T_g for these compounds with other parameters. The authors believe that to challenge these fundamental investigations, research on well-defined systems must become more effective over time, which can further the understanding of complicated active and biological systems. As such, the authors hope that this chapter inspires other researchers to initiate research relevant to the "glassy state and glass transition in amphiphilic lipids systems."

REFERENCES

Ahmad, Ramsch., Esquena, Solans, Tajuddin. & Rauzah, Hashim. (2012). "Physicochemical characterization of natural-like branched-chain glycosides toward formation of hexosomes and vesicles." *Langmuir*, *28*, 2395-2403.

Aloi, Oranges, Guzzi. & Rosa, Bartucci. (2017). "Low-Temperature Dynamics of Chain-Labeled Lipids in Ester-and Ether-Linked Phosphatidylcholine Membranes." *The Journal of Physical Chemistry B*, *121*, 9239-9246.

Amim Jr., Blachechen. & Denise F. S., Petri. (2012). "Effect of sorbitan-based surfactants on glass transition temperature of cellulose esters." *Journal of Thermal Analysis and Calorimetry*, *107*, 1259-1265.

Anderson, P. W. (1995). "Viewpoint: The Future." *Science*, *267*, 1615-1616.

Angell, C. A. (2002). "Liquid fragility and the glass transition in water and aqueous solutions." *Chemical Reviews*, *102*, 2627-2650.

Auzély-Velty, Benvegnu., Mackenzie, Haley, Goodby. & Daniel, Plusquellec. (1998a). "Synthesis and liquid-crystalline properties of novel archaeal diether-type glycolipids possessing one or two furanosyl units." *Carbohydrate Research*, *314*, 65-77.

Auzély-Velty, Benvegnu., Plusquellec, Mackenzie Haley. & John W., Goodby. (1998b). "Self-Organization and Formation of Liquid Crystal Phases by Molecular Templates Related to Membrane Components of

Archaebacteria." *Angewandte Chemie International Edition, 37*, 2511-2515.

Bancroft, W. D. (1913). "The theory of emulsification, V." *The Journal of Physical Chemistry, 17*, 501-519.

Berntsen, Svanberg. & J., Swenson. (2011). "Interplay between hydration water and headgroup dynamics in lipid bilayers." *The Journal of Physical Chemistry B, 115*, 1825-1832.

Blöcher, Gutermann, Henkel. & Klaus, Ring. (1984). "Physicochemical characterization of tetraether lipids from Thermoplasma acidophilum Differential scanning calorimetry studies on glycolipids and glycophospholipids." *Biochimica et Biophysica Acta (BBA)-Biomembranes, 778*, 74-80.

Blöcher, Six, Gutermann, Henkel. & Klaus, Ring. (1985). "Physicochemical characterization of tetraether lipids from Thermoplasma acidophilum. Calorimetric studies on miscibility with diether model lipids carrying branched or unbranched alkyl chains." *Biochimica et Biophysica Acta (BBA)-Biomembranes, 818*, 333-342.

Blöcher, Gutermann, Henkel. & Klaus, Ring. (1990). "Physicochemical characterization of tetraether lipids from Thermoplasma acidophilum. V. Evidence for the existence of a metastable state in lipids with acyclic hydrocarbon chains." *Biochimica et Biophysica Acta (BBA)-Biomembranes, 1024*, 54-60.

Boyd, Drummond, Krodkiewska. & Franz, Grieser. (2000). "How chain length, headgroup polymerization, and anomeric configuration govern the thermotropic and lyotropic liquid crystalline phase behavior and the air–water interfacial adsorption of glucose-based surfactants." *Langmuir, 16*, 7359-7367.

Brinke, Karasz. & Thomas, S., Ellis. (1983). "Depression of glass transition temperatures of polymer networks by diluents." *Macromolecules, 16*, 244-249.

Castellano, Generosi, Congiu. & R., Cantelli. (2006). "Glass transition temperature of water confined in lipid membranes as determined by anelastic spectroscopy." *Applied Physics Letters, 89*, 233905.

Chambert, Doutheau, Queneau, Cowling, Goodby, & Grahame, Mackenzie. (2007). "Synthesis and thermotropic behavior of simple new glucolipid amides." *Journal of Carbohydrate Chemistry*, *26*, 27-39.

Chen, Katsis., Schmid, Mastrangelo, Tsutsui. & T. N., Blanton. (1999). "Circularly polarized light generated by photoexcitation of luminophores in glassy liquid-crystal films." *Nature*, *397*, 506-508.

Chen, Ou. & Shaw, H. Chen. (2014). "Glassy Liquid Crystals as Self-Organized Films for Robust Optoelectronic Devices." In *Nanoscience with Liquid Crystals*, edited by Q. Li, 179-208. Springer: Cham. https://doi.org/10.1007/978-3-319-04867-3_6

Claesson, Kjellin, Rojas. & Cosima, Stubenrauch. (2006). "Short-range interactions between non-ionic surfactant layers." *Physical Chemistry Chemical Physics*, *8*, 5501-5514.

Clemente, Fitremann, Mauzac, Serrano. & Luis, Oriol. (2011). "Synthesis and characterization of maltose-based amphiphiles as supramolecular hydrogelators." *Langmuir*, *27*, 15236-15247.

Costello. & T., Gulik-Krzywicki. (1976). "Correlated X-ray diffraction and freeze-fracture studies on membrane model systems perturbations induced by freeze-fracture preparative procedures." *Biochimica et Biophysica Acta (BBA)-Biomembranes*, *455*, 412-432.

Crowe, Crowe, Carpenter, Rudolph., Wistrom, Spargo. & T. J., Anchordoguy (1988). "Interactions of sugars with membranes." *Biochimica et Biophysica Acta*, *947*, 367-384.

Couchman, P. R. (1978). "Compositional variation of glass-transition temperatures. 2. Application of the thermodynamic theory to compatible polymer blends." *Macromolecules*, *11*, 1156-1161.

Couchman, F. E. Karasz. (1978). "A classical thermodynamic discussion of the effect of composition on glass-transition temperatures." *Macromolecules*, *11*, 117-119.

Debenedetti. & Frank H., Stillinger. (2001). "Supercooled liquids and the glass transition." *Nature*, *410*, 259-267.

Dumoulin, Lafont., Boullanger, Mackenzie, Mehl. & John W., Goodby. (2002). "Self-organizing properties of natural and related synthetic glycolipids." *Journal of the American Chemical Society, 124*, 13737-13748.

Eastoe. & J. S., Dalton. (2000). "Dynamic surface tension and adsorption mechanisms of surfactants at the air–water interface." *Advances in Colloid and Interface Science, 85*, 103-144.

Ericsson, Ericsson., Kocherbitov, Söderman. & Stefan, Ulvenlund. (2005a). "Thermotropic phase behaviour of long-chain alkylmaltosides." *Physical Chemistry Chemical Physics, 7*, 2970-2977.

Ericsson, Ericsson. & Stefan, Ulvenlund. (2005b). "Solid-state phase behaviour of dodecylglycosides." *Carbohydrate research, 340*, 1529-1537.

Erilov, Bartucci., Guzzi, Marsh, Dzuba. & Luigi, Sportelli. (2004). "Librational motion of spin-labeled lipids in high-cholesterol containing membranes from echo-detected EPR spectra." *Biophysical Journal, 87*, 3873-3881.

Ernst, Freisleben, Antonopoulos, Henkel, Mlekusch. & Gilbert, Reibnegger. (1998). "Calorimetry of archaeal tetraether lipid—indication of a novel metastable thermotropic phase in the main phospholipid from Thermoplasma acidophilum cultured at 59°C." *Chemistry and physics of lipids, 94*, 1-12.

Fischer, Fischer., Diele, Pelzl., Jankowski, Schmidt. & Volkmer, Vill. (1994). "On the structure of the thermotropic cubic mesophases." *Liquid Crystals, 17*, 855-861.

Fong, Weerawardena., Sagnella, Mulet, Waddington, Krodkiewska. & Calum J., Drummond. (2010). "Monodisperse nonionic phytanyl ethylene oxide surfactants: high throughput lyotropic liquid crystalline phase determination and the formation of liposomes, hexosomes and cubosomes." *Soft Matter, 6*, 4727-4741.

Fong, Weerawardena, Sagnella, Mulet, Krodkiewska, Chong. & Calum J., Drummond. (2011). "Monodisperse nonionic isoprenoid-type hexahydrofarnesyl ethylene oxide surfactants: High throughput lyotropic liquid crystalline phase determination." *Langmuir, 27*, 2317-2326.

Fong, Wells, Krodkiewska, Weerawardeena, Booth, Hartley. & Calum J., Drummond. (2007). "Diversifying the solid state and lyotropic phase behavior of nonionic urea-based surfactants." *The Journal of Physical Chemistry B*, *111*, 10713-10722.

Gerber, Wulf, Milkereit, Vill, Howe, Roessle, Garidel, Gutsmannn. & K., Brandenbur, (2009). "Phase diagrams of monoacylated amide-linked disaccharide glycolipids." *Chemistry and Physics of Lipids*, *158*, 118-130.

Giri, Li Tuan. & Marcus, T., Cicerone. (2011). "Stabilization of proteins by nanoencapsulation in sugar-glass for tissue engineering and drug delivery applications." *Advanced Materials*, 23, 4861-4867.

Gliozzi, Paoli, Pisani, Gliozzi, De Rosa. & A., Gambacorta. (1986). "Phase transitions of bipolar lipids of thermophilic archaebacteria." *Biochimica et Biophysica Acta (BBA)-Biomembranes*, *861*, 420-428.

Gong, Moghaddam, Sagnella, Conn., Danon, Waddington. & Calum J., Drummond. (2011). "Lyotropic liquid crystalline self-assembly material behavior and nanoparticulate dispersions of a phytanyl pro-drug analogue of capecitabine– A chemotherapy agent." *ACS applied materials & interfaces*, *3*, 1552-1561.

Gong, Sagnella. & Caulm J., Drummond. (2008). "Nanostructured self-assembly materials formed by non-ionic urea amphiphiles." *International Journal of Nanotechnology*, *5*, 370-392.

Goodby, John. W. (1984). "Liquid crystal phases exhibited by some monosaccharides." *Molecular Crystals and Liquid Crystals*, *110*, 205-219.

Goodby, Görtz, Cowling, Mackenzie, Martin, Plusquellec., Benvegnu, Boullanger, Lafont, Queneau, Chambert. & J., Fitremann. (2007). "Thermotropic liquid crystalline glycolipids." *Chemical Society Reviews*, *36*, 1971-2032.

Gordon, Rouse, Gibbs. & W. M. Risen Jr. (1977). "The composition dependence of glass transition properties" *The Journal of Chemical Physics*, *66*, 4971-4976.

Häckl, Mühlbauer, Ontiveros, Marinkovic, Estrine, Kunz. & Véronique, Nardello-Rataj. (2018). "Carnitine alkyl ester bromides as novel biosourced ionic liquids, cationic hydrotropes and surfactants." *Journal of Colloid and Interface Science, 511,* 165-173.

Hashim, Zahid, Velayutham, Aripin, Ogawa. & Akihiko, Sugimura. (2018). "Dry thermotropic glycolipid self-assembly: a review." *Journal of Oleo Science, 67,* 651-668.

Hirsh, Allen. (1983). "An alternative explanation for phase transitions observed in quick frozen calcium cardiolipin solution." *Biochimica et Biophysica Acta (BBA)-Biomembranes, 733,* 186-188.

Hoffmann, Milius, Voss, Wunschel, van Smaalen, Diele. & Gerhard, Platz. (1999). "Crystal structures and thermotropic properties of alkyl α-D-glucopyranosides and their hydrates." *Carbohydrate research, 323,* 192-201.

Imamura, Murai, Korehisa, Shimizu, Yamahira, Matsuura, Tada, Imanaka Ishida. & Kazuhiro, Nakanishi. (2014). "Characteristics of sugar surfactants in stabilizing proteins during freeze-thawing and freeze-drying." *Journal of Pharmaceutical Sciences, 103,* 1628-1637.

Izutsu, Yoshioka. & Tadao, Terao. (1993). "Stabilization of β-galactosidase by amphiphilic additives during freeze-drying." *International Journal of Pharmaceutics, 90,* 187-194.

Izutsu, Yoshioka. & Tadao, Terao. (1994). "Stabilizing effect of amphiphilic excipients on the freeze-thawing and freeze-drying of lactate dehydrogenase." *Biotechnology and Bioengineering, 43,* 1102-1107.

Izutsu, Yoshioka. & Shigeo, Kojima. (1995). "Increased stabilizing effects of amphiphilic excipients on freeze-drying of lactate dehydrogenase (LDH) by dispersion into sugar matrices." *Pharmaceutical Research, 12,* 838-843.

Isaev, & Sergei A., Dzuba. (2008). "Fast stochastic librations and slow rotations of spin labeled stearic acids in a model phospholipid bilayer at cryogenic temperatures." *The Journal of Physical Chemistry B, 112,* 13285-13291.

Ishak, Zahid, Velayutham, Annuar. & Rauzah, Hashim. (2019). "Effects of lipid packing and intermolecular hydrogen bond on thermotropic phase transition of stearyl glucoside." *Journal of Molecular Liquids, 281,* 20-28.

Israelachvili, J. (1994). "Self-assembly in two dimensions: surface micelles and domain formation in monolayers." *Langmuir, 10*, 3774-3781.

Jana, Murthy, Jayaraman. & Channabasaveshwar V. Yelamaggad. (2006). "Aggregation and mesomorphic properties of 'double-headed' carbohydrate amphiphiles." *Phase Transitions, 78*, 529-535.

Jayaraman, Singh, Rao. & S. K., Prasad. (2007). "Studies of the mesomorphic behavior of bivalent carbohydrate amphiphiles." *Journal of Materials Chemistry, 17*, 2228-2232.

Jensen, O'Brien, Wang, Bryant, Ward, James. & D. A., Lewis. (1998). "Characterization of epoxy–surfactant interactions." *Journal of Polymer Science Part B: Polymer Physics, 36*, 2781-2792.

Júnior, Kawano, Denise. & F., Petri. (2009). "Thin films of carbohydrate based surfactants and carboxymethylcellulose acetate butyrate mixtures: Morphology and thermal behavior." *Materials Science and Engineering: C, 29*, 420-425.

Katkov. & Fred, Levine. (2004). "Prediction of the glass transition temperature of water solutions: comparison of different models." *Cryobiology, 49*, 62-82.

Kobayashi, Ichikawa, Kato. & Hiroyuki, Ohno. (2017). "Development of Glassy Bicontinuous Cubic Liquid Crystals for Solid Proton-Conductive Materials." *Advanced Materials, 29*, 1604429.

Kocherbitov. & & Olle, Söderman. (2003). "Phase diagram and physicochemical properties of the n-octyl α-D-glucoside/water system." *Physical Chemistry Chemical Physics, 5*, 5262-5270.

Kocherbitov. & Olle, Söderman. (2004). "Glassy crystalline state and water sorption of alkyl maltosides." *Langmuir, 20*, 3056-3061.

Kodama, Kuwabara. & Syûzo, Seki. (1981). "New Finding of the Glass Transition Phenomenon in Surfactant Gel Phase: The Binary System of Water and Dioctadecyldimethyl-Ammonium Chloride." *Molecular Crystals and Liquid Crystals, 64*, 277-282.

Koeltzow. & Allen, D. Urefer. (1984). "Preparation and properties of pure alkyl glucosides, maltosides and maltotriosides." *Journal of the American Oil Chemists' Society, 61*, 1651-1655.

Koster, Lei, Anderson, Martin. & Gray, Bryant. (2000). "Effects of vitrified and nonvitrified sugars on phosphatidylcholine fluid-to-gel phase transitions." *Biophysical Journal*, *78*, 1932-1946.

Kunitake, Toyoki. (1992). "Synthetic Bilayer Membranes: Molecular Design, Self-Organization, and Application." *Angewandte Chemie International Edition in English*, *31*, 709-726.

Langmuir, Irving. (1916). "The constitution and fundamental properties of solids and liquids. Part I. Solids." *Journal of the American Chemical Society*, *38*, 2221-2295.

Langmuir, Irving. (1917). "The constitution and fundamental properties of solids and liquids. II. Liquids." *Journal of the American Chemical Society*, *39*, 1848-1906.

Melchior, D. L. (1986). "Lipid domains in fluid membranes: a quick-freeze differential scanning calorimetry study." *Science*, *234*, 1577-1580.

Melchior, Bruggemann. & Joseph M., Steim. (1982). "The physical state of quick-frozen membranes and lipids." *Biochimica et Biophysica Acta (BBA)-Biomembranes*, *690*, 81-88.

Mackenzie, A. P. (1977). "Non-equilibrium freezing behaviour of aqueous systems." *Philosophical Transactions of the Royal Society of London. B, Biological Sciences*, *278*, 167-189.

Marcus, P. L. Finn. (1985). "The S_A phase of an alkylated sugar viewed as a lyotropic phase." *Molecular Crystals and Liquid Crystals Letters*, *2*, 159-166.

Merino, Rodrigues, Viciosa, Melo, Sotomayor, Dionísio. & Natália T., Correia. (2011). "Phase transformations undergone by triton X-100 probed by differential scanning calorimetry and dielectric relaxation spectroscopy." *The Journal of Physical Chemistry B*, *115*, 12336-12347.

Merino, Danéde, Derrollez, Dias, Viciosa, Correia. & Madalena, Dionísio. (2013a). "Investigating the Influence of Morphology in the Dynamical Behavior of Semicrystalline Triton X-100: Insights in the Detection/Nondetection of the α′-Process." *The Journal of Physical Chemistry B*, *117*, 9793-9805.

Merino, Neves, Fonseca, Danéde, Idrissi, Dias, Dionísio. & Natália T., Correia. (2013b). "Detection of two glass transitions on triton x-100 under confinement." *The Journal of Physical Chemistry C, 117,* 21516-21528.

Milkereit, Garamus, Veermans, Willumeit. & Volkmer, Vill. (2004). "Synthesis and mesogenic properties of a Y-shaped glyco-glycero-lipid." *Chemistry and physics of lipids, 131,* 51-61.

Milkereit, Gerber, Jankowski, Terjung, Schmidt. & Volkmer, Vill. (2005a). "From Y-To Siamese-twin Shaped Glycolipids: Influence on the thermotropic phase behaviour." *Journal of Thermal Analysis and Calorimetry, 82,* 471-475.

Milkereit, Garamus, Yamashita, Hato, Morr. & Volkmer, Vill. (2005b). "Comparison of the supramolecular structures of two glyco lipids with chiral and nonchiral methyl-branched alkyl chains from natural sources." *The Journal of Physical Chemistry B, 109,* 1599-1608.

Milkereit, Vasil. & Garamus, M. (2005). "Complex effect of ethyl branching on the supramolecular structure of a long chain neoglycolipid." *Colloids and Surfaces A: Physicochemical and Engineering Aspects, 268,* 155-161.

Miller, de Pablo. & Horacio, Corti. (1997). "Thermophysical properties of trehalose and its concentrated aqueous solutions." *Pharmaceutical Research, 14,* 578-590.

Moghaddam, de Campo, Waddington. & Calum J., Drummond. (2010). "Chelating phytanyl-EDTA amphiphiles: self-assembly and promise as contrast agents for medical imaging." *Soft Matter, 6,* 5915-5929.

Molinier, Kouwer, Fitremann, Bouchu, Mackenzie, Queneau. & John W., Goodby. (2006). "Self-organizing properties of monosubstituted sucrose fatty acid esters: the effects of chain length and unsaturation." *Chemistry–A European Journal, 12,* 3547-3557.

Nowacka, Bongartz, Ollila, Nylander. & D., Topgaard. (2013). "Signal intensities in ^1H–^{13}C CP and INEPT MAS NMR of liquid crystals." *Journal of Magnetic Resonance, 230,* 165-175.

Ogawa, Asakura. & Shuichi, Osanai. (2010). "Glass transition behavior of octyl *β*-D-glucoside and octyl *β*-D-thioglucoside/water binary mixtures." *Carbohydrate research, 345,* 2534-2541.

Ogawa, Asakura. & Shuchi, Osanai. (2012). "Freezing and melting behavior of an octyl β-D-glucoside–water binary system–inhibitory effect of octyl β-D-glucoside on ice crystal formation." *Physical Chemistry Chemical Physics*, *14*, 16312-16320.

Ogawa, Asakura. & Shuchi, Osanai. (2013). "Thermotropic and glass transition behaviors of n-alkyl β-D-glucosides." *RSC Advances*, *3*, 21439-21446.

Ogawa. & Osanai, S. (2012). "Glass transition behavior of aqueous solution of sugar-based surfactants." In *Supercooling*, edited by Peter Wilson, 27-55. IntechOpen.

Ogawa, Kawai, Koga, Asakura, Takahashi. & Shuichi, Osanai. (2015). "Degree of maintenance of lactate dehydrogenase (LDH) activity during freeze/thaw process highly depends on hydrophobic chain length of synthetic mono-tailed glycolipid stabilizer." *Cryobiology and Cryotechnology*, *61*, 89-94.

Ogawa, Shigesaburo. (2016). "Phase behavior of n-alkyl glucosides in the arid system—recent developments and experimental notes." In *Glucosides, sources, applications, and new research*, edited by Irving Boyd, 63-98. Nova Science Publisheers Inc: New York.

Ogawa, Kawai, Koga, Asakura, Takahashi. & Shuichi, Osanai. (2016). "Oligosaccharide-based surfactant/citric acid buffer system stabilizes lactate dehydrogenase during freeze-drying and storage without the addition of natural sugar." *Journal of Oleo Science*, *65*, 525-532.

Ogawa, Honda, Tsubomura, Totani, Takahashi. & Setuko, Hara. (2018). "Physicochemical characterization of 6-O-acyl trehalose fatty acid monoesters in desiccated system." *Chemistry and physics of lipids*, *216*, 80-90.

Patrick, Zahid, Kriechbaum. & Rauzah Hashim. (2018). "Guerbet glycolipids from mannose: liquid crystals properties." *Liquid Crystals*, *45*, 1970-1986.

Percec, Glodde, Bera, Miura, Shiyanovskaya, Singer, Balagurusamy, Heiney, Schnell, Rapp, Spiess, Hudson. & H., Duan. (2002). "Self-organization of supramolecular helical dendrimers into complex electronic materials." *Nature*, *419*, 384-387.

Perinelli, Lucarini, Fagioli, Campana, Vllasaliu, Duranti. & L. Casettari. (2018). "Lactose oleate as new biocompatible surfactant for pharmaceutical applications." *European Journal of Pharmaceutics and Biopharmaceutics, 124*, 55-62.

Pinal, Rodolfo. (2008). "Entropy of mixing and the glass transition of amorphous mixtures." *Entropy, 10*, 207-223.

Sackmann, Erich. (1968). "Polarization of optical transitions of dye molecules oriented in an ordered glass matrix." *Journal of the American Chemical Society, 90*, 3569-3570.

Sackmann, Meiboom, Snyder, Meixner. & R. E., Dietz. (1968). "Structure of the liquid crystalline state of cholesterol derivatives." *Journal of the American Chemical Society, 90*, 3567-3569.

Saari, Mislan, Hashim. & N. Zahid. (2018). "Self-Assembly, Thermotropic, and Lyotropic Phase Behavior of Guerbet Branched-Chain Maltosides." *Langmuir, 34*, 8962-8974.

Sagnella, Conn, Krodkiewska. & Calum J., Drummond. (2009). "Soft ordered mesoporous materials from nonionic isoprenoid-type monoethanolamide amphiphiles self-assembled in water." *Soft Matter, 5*, 4823-4834.

Sagnella, Conn, Krodkiewska. & Calum J., Drummond. (2011a). "Nonionic diethanolamide amphiphiles with isoprenoid-type hydrocarbon chains: thermotropic and lyotropic liquid crystalline phase behaviour." *Physical Chemistry Chemical Physics, 13*, 17511-17520.

Sagnella, Conn, Krodkiewska. & Calum J., Drummond. (2011b). "Nonionic diethanolamide amphiphiles with unsaturated C18 hydrocarbon chains: thermotropic and lyotropic liquid crystalline phase behavior." *Physical Chemistry Chemical Physics, 13*, 13370-13381.

Sagnella, Conn, Krodkiewska, Mulet. & Calum J., Drummond. (2011c). "Anandamide and analogous endocannabinoids: a lipid self-assembly study." *Soft Matter, 7*, 5319-5328.

Schmidt. & Karin, Jankowski. (1996). "New types of nonionic surfactants with sugar head groups." *Liebigs Annalen, 1996*, 867-879.

Shalaev. & Peter, L. Steponkus. (2001). "Phase behavior and glass transition of 1, 2-dioleoylphosphatidylethanolamine (DOPE) dehydrated in the presence of sucrose." *Biochimica et Biophysica Acta (BBA)-Biomembranes, 1514*, 100-116.

Shalaev. & Peter, L. Steponkus. (2003). "Glass transition of a synthetic phospholipid in the lamellar phase." *The Journal of Physical Chemistry B, 107*, 8734-8737.

Shalaev, Wu., Shamblin, Krzyzaniak. & Marc, Descamps. (2016). "Crystalline mesophases: structure, mobility, and pharmaceutical properties." *Advanced Drug Delivery Reviews, 100*, 194-211.

Shalaev, Zografi. & Peter, L. Steponkus. (2010). "Occurrence of glass transitions in long-chain phosphatidylcholine mesophases." *The Journal of Physical Chemistry B, 114*, 3526-3533.

Singh, & N., Jayaraman. (2009). "Carbohydrate-based liquid crystals." *Journal of the Indian Institute of Science, 89*, 113-135.

Singh, Jayaraman, Rao. & Krishna, Prasad. (2010). "Role of hydroxyl group on the mesomorphism of alkyl glycosides: synthesis and thermal behavior of alkyl 6-deoxy-β-D-glucopyranosides." *Chemistry and physics of lipids, 163*, 580-585.

Singh, Xu, Moebs, Kumar, Queneau, Cowling. & John W. Goodby. (2013). "Hydrophobic and hydrophilic balance and its effect on mesophase behaviour in hydroxyalkyl ethers of methyl glucopyranoside." *Chemistry–A European Journal, 19*, 5041-5049.

Slade, Levine. & D. S., Reid. (1991). "Beyond water activity: recent advances based on an alternative approach to the assessment of food quality and safety." *Critical Reviews in Food Science & Nutrition, 30*, 115-360.

Sorai, Michio. (1985). "Heat capacities of thermotropic liquid crystals with low molecular weight." *Thermochimica Acta, 88*, 1-16.

Sorai. & Syûzo, Seki. (1971). "Glassy liquid crystal of the nematic phase of *N*-(o-hydroxy-*p*-methoxybenzylidene)-p-butylaniline." *Bulletin of the Chemical Society of Japan, 44*, 2887-2887.

Sorai, Yoshioka. & Hiroshi, Suga. (1984). "Unusual glass transition of smectic liquid crystal in p-n-hexyloxybenzylidene-*p*'-butylaniline." In *Liquid Crystals and Ordered Fluids*, vol. *4*, edited by Anselm C. Griffin, 233-250. Springer US: Prenum Press, New York.

Suga. & Syûzo, Seki. (1980). "Frozen-in states of orientational and positional disorder in molecular solids." *Faraday Discussions of the Chemical Society*, *69*, 221-240.

Svanberg, Berntsen, Johansson, Hedlund. & Axén, J. Swenson. (2009). "Structural relaxations of phospholipids and water in planar membranes." *The Journal of Chemical Physics*, *130*, 01B609.

Swenson, Kargl, Berntsen. & C., Svanberg. (2008). "Solvent and lipid dynamics of hydrated lipid bilayers by incoherent quasielastic neutron scattering." The *Journal of Chemical Physics*, *129*, 07B616.

Szűts, Pallagi, Regdon Jr., Aigner. & Piroska, Szabó-Révész. (2007). "Study of thermal behaviour of sugar esters." *International journal of pharmaceutics*, *336*, 199-207.

Tiddy, G. J. (1980). "Surfactant-water liquid crystal phases." *Physics reports*, *57*, 1-46.

Tsuji, Sorai. & Syûzo, Seki. (1971). "New finding of glassy liquid crystal– a non-equilibrium state of cholesteryl hydrogen phthalate. *Bulletin of the Chemical Society of Japan.*," *44*, 1452-1452.

von Minden, Brandenburg, Seydel, Koch, Garamus, Willumeit. & Volkmar, Vill. (2000). "Thermotropic and lyotropic properties of long chain alkyl glycopyranosides. Part II. Disaccharide headgroups." *Chemistry and Physics of Lipids*, *106*, 157-179.

Voinova, M. V. (1995). "Phase transitions and glass-like behaviour of amphiphile assemblies." *Colloids and Surfaces A: Physicochemical and Engineering Aspects*, *95*, 133-139.

Voinova, M. V. (1996). "The theory of membrane "vitrification"." *Thermochimica acta*, *280*, 465-477.

Kauzmann, Walter. (1948). "The nature of the glassy state and the behavior of liquids at low temperatures." *Chemical reviews*, *43*, 219-256.

Wang. & Richard, M. Epand. (2004). "Factors determining pressure perturbation calorimetry measurements: evidence for the formation of metastable states at lipid phase transitions." *Chemistry and physics of lipids*, *129*, 21-30.

Wang, Velikov. & C. Austen, Angell. (2002). "Direct determination of kinetic fragility indices of glass forming liquids by differential scanning calorimetry: Kinetic versus thermodynamic fragilities." *The Journal of chemical physics*, *117*, 10184-10192.

Wang, Angell. & Ranko, Richert. (2006). "Fragility and thermodynamics in nonpolymeric glass-forming liquids." *The Journal of Chemical Physics*, *125*, 074505.

Winsor, P. A. (1968). "Binary and multicomponent solutions of amphiphilic compounds. Solubilization and the formation, structure, and theoretical significance of liquid crystalline solutions." *Chemical Reviews*, *68*, 1-40.

Biographical Sketch

Shigesaburo Ogawa

Affiliation: Department of Materials and Life Science, Faculty of Science and Technology, Seikei University

Education:

2004	BS Applied Chemistry, Keio University
2006	MEng, Integrated Design Engineering, Keio University
2010	PhD (Dr. Eng.), School of Fundamental Science and Technology, Keio University

Doctoral Dissertation: Study on the behaviors of sugar-based surfactant and water mixture system under low-temperature (in Japanese)

Research and Professional Experience: Dr. Ogawa has experience in both organic synthesis and materials analysis. He has conducted research on lipids, surfactants, oils, sugars, vitamins, dyes, polymers, and so on. His research is mainly based on surface chemistry, organic chemistry, applied chemistry, food science, green chemistry, and glassy materials.

Until March 2010, he mainly engaged in the synthesis of glycolipids and characterization of their aqueous solutions (Gibbs adsorption and micelles), coagels, and lyotropic and thermotropic liquid crystals (LCs) and their corresponding glassy state. These studies focused on the behavior, in particular, in the freeze–thawing process, and the application of synthetic glycolipid compounds for stabilizing therapeutic protein in the freeze–thawing and freeze–drying process. Meanwhile, he also engaged in several investigations related to the ice-generating system (e.g., ice nucleating emulsion systems and ice coalescing system). In his studies on protein stabilization using synthetic glycolipids and ice nucleation systems, he was assisted by two junior students (Maito Koga from April 2008 to March 2010 and Ryuichiro Kawai from April 2009 to March 2010, from the Oleo Science Laboratory, Keio University, supervised by Professor Osanai) in the experiments.

Except for the characterization of aqueous surfactant solution in terms of micelle formation and Gibbs adsorption film, which had been conventionally studied in the Oleo Science Laboratory, he was directly associated with the launching of the studies mentioned in the previous paragraph. In the works that he joined midway, the initial motivation was based on a discovery by senior students. But, his research direction was different from that of the previous works by senior students of the Oleo Science Laboratory. The senior students had reported that an interesting phenomenon (large exothermic peak) appeared in the heating process of dodecyl raffinose isomeric mixtures-water systems after rapid cooling; these mixtures are prepared by a one-step reaction between raffinose and dodecyl bromide in the dimethylformaide (DMF) with the addition of NaH via Williamson reaction. However, Dr. Ogawa proved that highly purified dodecyl raffinose compounds did not exhibit such behavior at all. Instead, he found that the addition of such glycolipids to aqueous electrolytes solutions can prevent precipitation of the eutectic mixture consisting of electrolyte and ice. So, from March 2005, he started research to understand the interaction of the sugar moiety of glycolipids and electrolytes in the freezing state. During the process, he noticed that the typical thermal behavior of glycolipid-electrolyte-water systems in the freeze–thawing processes seemed to be similar to those for the (non-amphiphilic) natural sugar-electrolyte-water system, though the research using glycolipids further required him to consider the formation of the LC phases. As he gradually understood the previous research related to the (non-amphiphilic) natural sugar-electrolyte-water system, he regarded it as important to consider the freeze-concentration effect on the non-frozen phase, where the concentrated glycolipid-water matrix was assumed to form lyotropic LC phase, even though the initial state prior to freezing was micelle solution. While the other members in the laboratory had eagerly studied the characterization of liposome and gel-LC phase transition for water insoluble phospholipid compounds, no seniors had studied both lyotropic LC phase and thermotropic LC phase of water soluble glycolipids. Therefore, the establishment of these relevant studies had to be carried out simultaneously. Among such developments, he discovered that most of the glycolipid compounds, even alkyl β-D-glucosides, whose glass-forming abilities were denied by Kocherbitov and Söderman in 2004, could be glass-formers. Based on these initial findings, the detailed investigations on glass transition of glycolipid have distinctly started in his research career; that is, no idea associated with glass transition existed initially, though the glass transition of glycolipids became a central research target in his doctoral dissertation. No description of the glass transition for a surfactant system is available in Japanese literature close to him, and it was less probable that he could encounter the concept of glass transition from his original background.

Professor Osanai was the supervisor in the Oleo Science Laboratory and often encouraged the students to equip themselves with the "spirit of independence and self-respect." Some students, at times, gained the opportunity to propose the research themes, even just after entering the laboratory. Professor Osanai often said "let's study something interesting," and this sentence has greatly contributed to the establishment of Dr. Ogawa's basic principle of early research outlook. Without such a mind for research and the permission from Professor Osanai, even half of his Doctoral dissertation would not have been conducted. The growing of research outlook of Dr. Ogawa was also affected by Associated Professor Kouichi Asakura (Professor of Keio University from 2009), in particular, in his early days.

Alternatively, the research was greatly supported by the kindness of Professor Shuichi Matsumura (Eminent Professor of Keio University from 2012), who gave him the opportunity to perform differential scanning calorimetry (DSC), thermogravimetric analysis, and mass spectroscopy analysis. Without his kindness, Dr. Ogawa's research approach would have been totally different. Dr. Ogawa had engaged in DSC measurement clocking over 5,000 hours in Keio University until March 2010.

Between April 2010 and March 2013, he was involved in a variety of research under the supervision of Professor Nobuo Kimizuka and Assistant Professor Masa-aki Morikawa. Dr. Ogawa had tried to develop a novel high performance ferroelectric material made of organic materials, hybrid, and water. Along with them, he developed a luminescent material consisting of synthetically prepared novel lipophilic boron difluoride (BF_2) complex compounds. The aggregation behaviors in these compounds in solution, concentrated forms, bulk crystals, and thin films were studied in detail. Taking into consideration the individual characteristic such as polymorphism in solid form of a variety of synthesized compounds, he developed a dual emissive material and stimuli-responsive molecular organized film systems that exhibit mechanochromism, vapochromism, cronochromism, turn-on emission, and so on. Additionally, a high emissive supramolecular gel system was developed via solvophobic interactions. Some of these works were not published but have been reported in domestic conferences; research in these works is undergoing. Some parts of these researches were advanced along with fruitful discussion with Dr. Takao Noguchi and appropriate supports by Dr. Juhasz Gergely.

Between April 2013 and March 2014, he engaged in research related to liquid-in-air dispersing systems, so-called liquid marble, in which the liquid droplet surface was encapsulated by monodisperse poly(methylsilsesquioxane) (PMSQ) particles. He developed the facile fabrication procedure to eliminate the excess adsorbed particles. As a consequence, he successfully discovered that the particular layer of these liquid marbles could be changed from multiple to single layered as the surface tension of internal liquids increased. Confocal laser scanning microscopy (CLSM) clearly showed the variation in particulate layers. In addition to the research, he engaged in the preparation and characterization of PMSQ films (e.g., wetting behavior), and so on. These works were supervised by Professor Atsushi Takahara, Professor Thomas J. McCarthy, Professor Hiroshi Jinnai, and Associate Professor Hirohmi Watanabe. Also, some of these works were performed with appropriate supports by other Ph.D. researchers (e.g., Dr. Liming Wang, Dr. Jin Nishida, and Dr. Yuki Norizoe) and Aya Fujimoto. Investigations of these works are still undergoing.

Between April 2014 and March 2016, he successfully applied the grazing-incidence two-dimensional wide angle X-ray diffraction (2D-GI-WAXD) analysis to discover the unknown crystal structure and phase behavior under the desiccated condition of mono-tailed glycolipids along with the establishment of the self-organization and characterization of sugar nanofilms consisting of cyclodextrin. These works were supervised by Professor Isao Takahashi. Self-organization of nano-sheets and multiple spiral nanofibers were successfully achieved using non-specifically modified cyclodextrin compounds in the absence of an organic solvent and a polymer. Dr. Ogawa also launched an investigation on the glass transition of glycolipid LC systems using thin films; Yoshitaka Ono also engaged its launch. Dr. Ogawa also engaged the study of the glass transition of polymer and sugar in thin films that had been already launched by Professor Isao Takahashi in the laboratory. Dr. Ogawa, in particular, contributed to establish the study, "determination of glass transition of cyclic sugar nanofilms by X-ray reflectivity." Research is still continuing on these aspects.

From April 2016 to March 2019, in addition to the glycolipid-related research, he engaged in research on oily materials (consisting of triacylglycerol, fatty acid, fatty acid ester, phospholipid, phenolic lipid, vitamin, etc.) in terms of enzymatic modifications, antioxidative properties, and their aggregation property under the supervisor of Professor Setsuko Hara (She retired in March 2018 and become Visiting Professor). These experiments were engaged together with many students. In the glycolipid study, the focus is on investigations on solvent-free enzymatic synthesis, together with the characterization of trehalose fatty acid ester in a desiccated system. The lamellar gel (L_β)-fluid lamellar (L_α) phase transition in the glassy state was firstly discovered for the glycolipid study to the best of his knowledge. He is also engaging in the characterization of antioxidative vitamin E compounds in water system in both presence and absence of cyclodextrin stabilizer.

From April 2019, he will start research on organic-hybrid luminescent material under the supervision of Professor Taro Tsubomura. Together with continuing his previous works, Dr. Ogawa intends to develop novel glassy luminescence materials.

Professional Appointments:

04/2010–03/2012   Research Assistant Professor in Kyushu University (Prof. Nobuo Kimizuka's group)

04/2012–03/2013   Post Doctoral Researcher in Kyushu University (Prof. Nobuo Kimizuka's group)

04/2013–03/2014   Research Assistant Professor in Japan Science and Technology Agency (JST), ERATO Takahara Soft Interfaces Project (Prof. Atsushi Takahara's group)

04/2014–03/2016   Post Doctoral Researcher in Kwansei Gakuin University (Prof. Isao Takahashi's group)

04/2016–03/2019   Assistant Professor in Seikei University (Prof. Setsuko Hara's group)

04/2019–   Assistant Professor in Seikei University (Prof. Taro Tsubomura's group)

Publications from the Last 3 Years:

Ogawa, Ashida, Kaneko. & Isao, Takahashi. (2018). "Self-organisation and characterisation of hierarchical structures in trimethyl β-cyclodextrin nano-films." *Materials Chemistry Frontiers*, *2*, 2191-2200.

Ogawa, Honda, Tsubomura, Totani, Takahashi. & Setsuko, Hara. (2018). "Physicochemical characterization of 6-*O*-acyl trehalose fatty acid monoesters in desiccated system." *Chemistry and Physics of Lipids*, *216*, 80-90.

Hashim, Zahid, Velayutham, Aripin, Ogawa. & Akihiko, Sugimura. (2018). "Dry thermotropic glycolipid self-assembly: A review." *Journal of Oleo Science*, *67*, 651-668.

Ogawa, Takahashi, Koga, Asakura. & Shuichi, Osanai. (2018). "Effect of freeze–thaw treatment on the precipitation of octyl *β*-D-galactoside hemihydrate crystal from the aqueous solution." *Journal of Oleo Science*, *67*, 627-637.

Ogawa. & Isao, Takahashi. (2018). "Temperature-dependent thickness variation of ultrathin trimethyl *β*-cyclodextrin film supported by a Si substrate." *Cryobiology and Cryotechnology*, *64*, 19-27.

Ogawa. & Isao, Takahashi. (2017). "Structural characterization of perpendicularly aligned submicrometer-thick synthetic glycolipid polycrystalline films using conventional X-ray diffraction." *Crystals*, *7*, 356.

Ogawa, Asakura. & Shuichi, Osanai. (2017). "Identification of a relationship between solute concentration and the temperature at which ice is well dispersed during freezing and thawing processes of aqueous low-molecular-weight-compound systems." *Cryobiology and Cryo-technology*, *63*, 95-102.

Ogawa, Koga, Kouichi, Asakura, Takahashi. & Shuichi, Osanai. (2017). "Coagel prepared from aqueous octyl *β*-D-galactoside solution." *Journal of Surfactants and Detergents*, *20*, 255-261.

Ogawa. & Isao, Takahashi. (2017). "Glass transition of ultrathin sugar films probed by X-ray reflectivity" In *Carbohydrate* edited by Mahmut Caliskan, Halil Kavakli, and Gül Cevahir Öz, 115-130. InTech.

Ogawa, Shigesaburo. (2016). "Phase behavior of n-alkyl glucosides in the arid system—Recent developments and experiment notes" In *Glucosides: Sources, Applications, and New Research* edited by Boyd, Irving, 63-98. Nova Science Publishers Inc.

Hoshino, Nojima, Sato, Hirai, Higaki, Fujinami, Murakami, Ogawa, Jinnai. & Atsushi, Takahara. (2016). "Observation of constraint surface dynamics of polystyrene thin films by functionalization of a silsesquioxane cage." *Polymer*, *105*, 487-499.

Ogawa, Ozaki. & Isao, Takahashi. (2016). "Structural insights into solid-to-solid phase transition and modulated crystal formation in octyl-*β*-D-galactoside crystals." *ChemPhysChem, 17*, 2808-2812.

Khasanah, Raghunatha Reddy, Ogawa, Sato, Takahashi. & Yukihiro, Ozaki. (2016). "Evolution of Intermediate and highly ordered crystalline states under spatial confinement in poly(3-hydroxybutyrate) ultrathin films." *Macromolecules, 49*, 4202-4210.

Ogawa, Kawai, Koga, Asakura, Takahashi. & Shuichi, Osanai. (2016). "Oligosaccharide-based surfactant/citric acid buffer system stabilizes lactate dehydrogenase during freeze-drying and storage without the addition of natural sugar." *Journal of Oleo Science, 65*, 525-532.

INDEX

H

I

U

V

W

X

Y

β

Δ
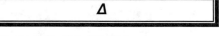

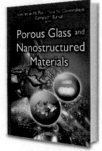